Quantum
Communications
and
Cryptography

Quantum Communications and Cryptography

edited by
Alexander V. Sergienko

CRC Press
Taylor & Francis Group
Boca Raton London New York

CRC Press is an imprint of the
Taylor & Francis Group, an **informa** business
A TAYLOR & FRANCIS BOOK

CRC Press
Taylor & Francis Group
6000 Broken Sound Parkway NW, Suite 300
Boca Raton, FL 33487-2742

First issued in paperback 2019

ISBN-13: 978-0-8493-3684-3 (hbk)
ISBN-13: 978-0-367-39174-4 (pbk)

Library of Congress Card Number 2005050636

Library of Congress Cataloging-in-Publication Data

Quantum communications and cryptography / Alexander V. Sergienko, editor.
     p. cm.
  Includes bibliographical references and index.
  ISBN 0-8493-3684-8
    1. Quantum communication--Security measures. 2. Cryptography. 3. Coding theory. 4. Data encryption (Computer science) I. Sergienko, Alexander V.

TK5102.94.Q36 2005
005.8--dc22
                                                       2005050636

Visit the Taylor & Francis Web site at
http://www.taylorandfrancis.com

and the CRC Press Web site at
http://www.crcpress.com

Preface

The amount of Internet traffic transmitted over optical telecommunication networks has seen an enormous surge over the last decade. This process is likely to continue considering the demand for a greater variety of services and faster download rates. One central issue of modern optical telecommunication is its security. Current communication security protection schemes are based on the mathematical complexity of specific encoding protocols. Any of them can, in principle, be deciphered when a sufficient computational power becomes available. There exists one particular scheme that is not vulnerable to such scenario — the one-time pad protocol. It is based on the condition of sharing secret random key material between two parties and using it for encrypting their information exchange. However, such random key material can be used only once and then must be discarded to ensure absolute security. This requires the key to be constantly refilled in such a way that only two legitimate users will possess identical sets of random key numbers. It is of the utmost importance to make sure that nobody else has gained access to the key material during refill procedures. This is where the use of special properties of the quantum state of light — the photon — offers a solution to the problem. Such basic principles of quantum theory as the no-cloning theorem have enabled researchers to implement a totally secure quantum key distribution (QKD). Secure distribution of random key material using quantum state of light constitutes the essence of a recently emerged area of physics and technology — quantum cryptography.

In 2005, quantum mechanics and quantum theory of light celebrated their 100th anniversary of successfully describing basic properties of matter and its interaction with electromagnetic radiation. Basic quantum principles outlined in earlier days have paved the way for the development of novel techniques for information manipulation that is based on the physical principles of correlation, superposition, and entanglement. Quantum information processing uses nonclassical properties of a quantum system in a superposition state (qubit) as the physical carrier of information. This is in contrast with conventional description, which is based on the use of discrete classical deterministic bits. This nonclassical manipulation of information has created the possibility of constructing extremely efficient quantum computers operating on thousands of qubits at a time. This challenging and far-reaching goal still requires a great deal of theoretical and experimental research efforts to develop quantum hardware resistant to decoherence and designing novel algorithms to serve as quantum software.

In the meantime, quantum information processing applications dealing with only a few qubits have been developed during the last decade and have been moving from the university and government research labs into the area of industrial research and development. Quantum cryptography that is based on the use of only one or two qubits can serve as a success story of practical quantum information processing. Several small businesses have already started offering practical point-to-point quantum key distribution devices covering short and medium distances thus developing a novel market for this disruptive technology. The first public quantum key distribution network that connects multiple users over commercial fibers in a metropolitan area has been operational for more than a year. Its constant development and expansion creates a solid foundation for heterogeneous architecture similar to the initial stages of Internet development.

This book aims at delivering a general overview of scientific foundations, theoretical and experimental results, and specific technological and engineering developments in quantum communication and cryptography demonstrated to date in university and government research laboratories around the world. The book is intended to serve as an introduction to the area of quantum information and, in particular, quantum communication and cryptography. The book is oriented towards graduate students in physics and engineering programs, research scientists, telecommunication engineers, and anybody who is enthusiastic about the power of quantum mechanics and who would be excited to learn about the emerging area of quantum optical communication.

The book opens with a brief history of conventional communication encoding and the appearance of quantum cryptography. Several fascinating experiments illustrating quantum information processing with entangled photons ranging from long-distance quantum key distribution in fiber to quantum teleportation of unknown state of light have been presented. These research efforts set a solid foundation for practical use of optical entanglement in quantum communication. Long-distance open-air quantum key distribution experiments have demonstrated the feasibility of extending quantum communication from the ground to a satellite and in between satellites in free space. The architecture of a currently operational metropolitan QKD network serving as the first heterogeneous quantum cryptography test-bed is described in detail. It is followed by the detailed theoretical analysis of practically meaningful security bounds. Several quantum communication protocols using continuous variables of nonclassical states of light are also presented. More complex applications of entangled states with few optical qubits are also described establishing building blocks for constructing linear-optical quantum computers and developing schemes for noise-immune quantum communications. This book was written by a group of physicists, engineers, and industrial scientists who are recognized leaders in the field of practical quantum information processing and quantum communication. References provided at the end of each chapter could be used as a guide for more detailed investigation of specific technical and scientific problems associated with this rapidly growing and very exciting area of science and technology.

I hope you enjoy reading the book.

Alexander V. Sergienko

Editor

Alexander V. Sergienko (e-mail: *alexserg@bu.edu*; URL: *http://people.bu.edu/alexserg*) received his M.S. and Ph.D. degrees in physics from Moscow State University in 1981 and 1987, respectively. After spending the 1990–1991 academic year at the University of Maryland College Park as a visiting professor, he joined the University of Maryland Baltimore County as a research assistant professor in 1991. He was associated with the National Institute of Standards and Technology (NIST) in Gaithersburg, Maryland, as a guest researcher from 1992 to 1996.

In 1996, Professor Sergienko joined the faculty of the Department of Electrical and Computer Engineering at Boston University. He holds joint appointments in the Department of Electrical and Computer Engineering and in the Department of Physics. He is also a codirector of the Quantum Imaging Laboratory at Boston University. His research interests include quantum information processing including quantum cryptography and communications, quantum imaging, the development of novel optical-measurement and characterization techniques based on the use of nonclassical states of light (quantum metrology), the experimental study of the basic concepts of quantum mechanics, the study of fundamental optical interactions of light with matter including quantum surface effects, and ultrafast quantum optics. He pioneered the experimental development of practical quantum-measurement techniques using entangled-photon states in the early 1980s.

Professor Sergienko has published more than 200 research papers in the area of experimental nonlinear and quantum optics. He holds five patents in the fields of nonlinear and quantum optics. He is a fellow of the Optical Society of America, a member of the American Physical Society, and a member of the IEEE\LEOS.

Contributors

Markus Aspelmeyer
Institute for Experimental Physics
University of Vienna
Vienna, Austria

Institute for Quantum Optics and
    Quantum Information
Austrian Academy of Sciences
Vienna, Austria

Hannes R. Böhm
Institute for Experimental Physics
University of Vienna
Vienna, Austria

Warwick P. Bowen
Department of Physics
Australian National University
Canberra, Australia

Artur Ekert
Department of Applied
    Mathematics and Theoretical
    Physics
University of Cambridge
Cambridge, United Kingdom

Chip Elliott
BBN Technologies
Cambridge, Massachusetts

Alessandro Fedrizzi
Institute for Experimental
    Physics
University of Vienna
Vienna, Austria

James D. Franson
Applied Physics Laboratory
Johns Hopkins University
Laurel, Maryland

Sara Gasparoni
Institute for Experimental Physics
University of Vienna
Vienna, Austria

Gerald Gilbert
Quantum Information Science
    Group, MITRE
Eatontown, New Jersey

Nicolas Gisin
Group of Applied Physics
University of Geneva
Geneva, Switzerland

P.M. Gorman
QinetiQ
Malvern, United Kingdom

M. Halder
Ludwig Maximilians University
    Munich
Munich, Germany

Group of Applied Physics
University of Geneva
Geneva, Switzerland

M. Hamrick
Quantum Information Science
    Group, MITRE
Eatontown, New Jersey

S. Iblisdir
Group of Applied Physics
University of Geneva
Geneva, Switzerland

B.C. Jacobs
Applied Physics Laboratory
Johns Hopkins University
Laurel, Maryland

Thomas D. Jennewein
Institute for Quantum Optics and
    Quantum Information
Austrian Academy of Sciences
Vienna, Austria

Natalia Korolkova
Friedrich Alexander University
    of Erlangen-Nürnberg
Erlangen, Germany

School of Physics and Astronomy
University of St. Andrews
St. Andrews, Scotland

Christian Kurtsiefer
Ludwig Maximilians University
    Munich
Munich, Germany

National University of Singapore
Singapore

Ping Koy Lam
Department of Physics
Australian National University
Canberra, Australia

Andrew Matheson Lance
Department of Physics
Australian National University
Canberra, Australia

Gerd Leuchs
Friedrich Alexander University
    of Erlangen-Nürnberg
Erlangen, Germany

Michael Lindenthal
Institute for Experimental Physics
University of Vienna
Vienna, Austria

S. Lorenz
Friedrich Alexander University
    of Erlangen-Nürnberg
Erlangen, Germany

N. Lütkenhaus
Friedrich Alexander University
    of Erlangen-Nürnberg
Erlangen, Germany

Gabriel Molina-Terriza
Institute for Experimental Physics
University of Vienna
Vienna, Austria

T.B. Pittman
Applied Physics Laboratory
Johns Hopkins University
Laurel, Maryland

Andreas Poppe
Institute for Experimental Physics
University of Vienna
Vienna, Austria

Timothy C. Ralph
Department of Physics
University of Queensland
Brisbane, Australia

John G. Rarity
Department of Electrical and
    Electronic Engineering
University of Bristol
Bristol, United Kingdom

Kevin Resch
Institute for Experimental Physics
University of Vienna
Vienna, Austria

Bahaa E.A. Saleh
Quantum Imaging Laboratory
Department of Electrical and
    Computer Engineering
Department of Physics
Boston University
Boston, Massachusetts

Barry C. Sanders
Department of Physics
  and Astronomy
University of Calgary
Calgary, Alberta, Canada

Alexander V. Sergienko
Quantum Imaging Laboratory
Department of Electrical and
  Computer Engineering
Department of Physics
Boston University
Boston, Massachusetts

Thomas Symul
Department of Physics
Australian National
  University
Canberra, Australia

P.R. Tapster
QinetiQ
Malvern, United Kingdom

Malvin C. Teich
Quantum Imaging Laboratory
Department of Electrical and
  Computer Engineering
Department of Physics
Boston University
Boston, Massachusetts

F.J. Thayer
Quantum Information Science
  Group, MITRE
Eatontown, New Jersey

W. Tittel
Group of Applied Physics
University of Geneva
Geneva, Switzerland

Rupert Ursin
Institute for Experimental
  Physics
University of Vienna
Vienna, Austria

Philip Walther
Institute for Experimental
  Physics
University of Vienna
Vienna, Austria

Zachary D. Walton
Quantum Imaging Laboratory
Department of Electrical and
  Computer Engineering
Department of Physics
Boston University
Boston, Massachusetts

Harald Weinfurter
Ludwig Maximilians University
  Munich
Munich, Germany

Max-Planck-Institute for Quantum
  Optics
Garching, Germany

P. Zarda
Ludwig Maximilians University
  Munich
Munich, Germany

Max-Planck-Institute for Quantum
  Optics
Garching, Germany

H. Zbinden
Group of Applied Physics
University of Geneva
Geneva, Switzerland

Anton Zeilinger
Institute for Experimental Physics
University of Vienna
Vienna, Austria

Institute for Quantum Optics and
  Quantum Information
Austrian Academy of Sciences
Vienna, Austria

Contents

Chapter 1   Quantum Cryptography 1
A. Ekert

Chapter 2   Quantum Communications with Optical Fibers 17
N. Gisin, S. Iblisdir, W. Tittel, and H. Zbinden

**Chapter 3   Advanced Quantum Communications Experiments
with Entangled Photons** 45
*M. Aspelmeyer, H. R. Böhm, A. Fedrizzi, S. Gasparoni, M. Lindenthal, G. Molina-
Terriza, A. Poppe, K. Resch, R. Ursin, P. Walther, A. Zeilinger, and T. D. Jennewein*

Chapter 4   The DARPA Quantum Network 83
C. Elliott

**Chapter 5   Experimental Cryptography Using Continuous
Polarization States** 103
S. Lorenz, N. Lütkenhaus, G. Leuchs, and N. Korolkova

Chapter 6   Quantum Logic Using Linear Optics 127
J.D. Franson, B.C. Jacobs, and T.B. Pittman

**Chapter 7   Practical Quantum Cryptography: Secrecy Capacity
and Privacy Amplification** 145
G. Gilbert, M. Hamrick, and F.J. Thayer

Chapter 8   Quantum State Sharing 163
T. Symul, A.M. Lance, W.P. Bowen, P.K. Lam, B.C. Sanders, and T.C. Ralph

Chapter 9   Free-Space Quantum Cryptography 187
C. Kurtsiefer, M. Halder, H. Weinfurter, P. Zarda, P.R. Tapster,
P.M. Gorman, and J.G. Rarity

Chapter 10   Noise-Immune Quantum Key Distribution 211
Z.D. Walton, A.V. Sergienko, B.E.A. Saleh, and M.C. Teich

Index 225

chapter 1

Quantum Cryptography

A. Ekert
University of Cambridge

Contents

1.1   Classical Origins 2
1.2   Le Chiffre Indéchiffrable 3
1.3   Not So Unbreakable.. 4
1.4   Truly Unbreakable? 5
1.5   Key Distribution Problem... 6
1.6   Local Realism and Eavesdropping..................................... 8
1.7   Quantum Key Distribution 9
      1.7.1   Entanglement-Based Protocols 9
      1.7.2   Prepare and Measure Protocols 10
1.8   Security Proofs 11
1.9   Concluding Remarks 13
References 13

Few persons can be made to believe that it is not quite an easy thing to invent a method of secret writing which shall baffle investigation. Yet it may be roundly asserted that human ingenuity cannot concoct a cipher which human ingenuity cannot resolve...

— Edgar Alan Poe, "A Few Words on Secret Writing," 1841

Abstract

Quantum cryptography offers new methods of secure communication. Unlike traditional classical cryptography, which employs various mathematical techniques to restrict eavesdroppers from learning the contents of encrypted messages, quantum cryptography is focused on the physics of information. The process of sending and storing information is always carried out by physical means, for example photons in optical fibers or electrons in electric current. Eavesdropping can be viewed as measurements on a physical object — in this case the carrier of the information. What the eavesdropper can measure, and how, depends exclusively on the laws of physics. Using quantum phenomena, we can design and implement a communication system that can always detect eavesdropping. This is because measurements on the quantum carrier of information disturb it and so leave traces. What follows is a brief overview of the quest for constructing unbreakable ciphers, from classical to quantum.

1.1 Classical Origins

Human desire to communicate secretly is at least as old as writing itself and goes back to the beginnings of civilization. Methods of secret communication were developed by many ancient societies, including those of Mesopotamia, Egypt, India, China, and Japan, but details regarding the origins of cryptology, i.e., the science and art of secure communication, remain unknown.

We know that it was the Spartans, the most warlike of the Greeks, who pioneered cryptography in Europe. Around 400 B.C. they employed a device known as the scytale (Figure 1.1). The device, used for communication between military commanders, consisted of a tapered baton around which was wrapped a spiral strip of parchment or leather containing the message. Words were then written lengthwise along the baton, one letter on each revolution of the strip. When unwrapped, the letters of the message appeared scrambled

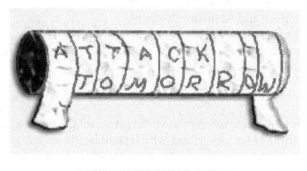

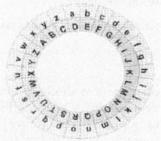

Figure 1.1 Scytale (top) and Alberti's disk (bottom) were the first cryptographic devices implementing permutations and substitutions, respectively.

and the parchment was sent on its way. The receiver wrapped the parchment around another baton of the same shape and the original message reappeared.

In his correspondence, Julius Caesar allegedly used a simple letter substitution method. Each letter of Caesar's message was replaced by the letter that followed it alphabetically by three places. The letter A was replaced by D, the letter B by E, and so on. For example, the English word COLD after the Caesar substitution appears as FROG. This method is still called the Caesar cipher, regardless of the size of the shift used for the substitution.

These two simple examples already contain the two basic methods of encryption which are still employed by cryptographers today, namely, *transposition* and *substitution*. In transposition (scytale) the letters of the *plaintext*, the technical term for the message to be transmitted, are rearranged by a special permutation. In substitution (Caesar's cipher) the letters of the plaintext are replaced by other letters, numbers or arbitrary symbols. The two techniques can be combined to produce more complex ciphers.

Simple substitution ciphers are easy to break. For example, the Caesar cipher with 25 letters admits any shift between 1 and 25, so it has 25 possible substitutions (or 26 if you allow the zero shift). One can easily try them all, one by one. The most general form of one-to-one substitution, not restricted to the shifts, can generate

$$26! \quad \text{or} \quad 403,291,461,126,605,635,584,000,000 \quad (1.1)$$

possible substitutions. And yet, ciphers based on one-to-one substitutions, also known as monoalphabetic ciphers, can be easily broken by frequency analysis. The method was proposed by the ninth-century polymath from Baghdad, Al-Kindi (800–873 A.D.), often called the philosopher of the Arabs.

Al-Kindi noticed that if a letter in a message is replaced with a different letter or symbol then the new letter will take on all the characteristics of the original one. A simple substitution cipher cannot disguise certain features of the message, such as the relative frequencies of the different characters. Take the English language: the letter E is the most common letter, accounting for 12.7% of all letters, followed by T (9.0%), then A (8.2%) and so on. This means that if E is replaced by a symbol X, then X will account for roughly 13% of symbols in the concealed message, thus one can work out that X actually represents E. Then we look for the second most frequent character in the concealed message and identify it with the letter T, and so on. If the concealed message is sufficiently long then it is possible to reveal its content simply by analyzing the frequency of the characters.

1.2 Le Chiffre Indéchiffrable

In the fifteenth and sixteenth centuries, monoalphabetic ciphers were gradually replaced by more sophisticated methods. At the time, Europe, Italy in particular, was a place of turmoil, intrigue, and struggle for political and financial power, and the cloak-and-dagger atmosphere was ideal for cryptography to flourish.

In the 1460s Leone Battista Alberti (1404–1472), better known as an architect, invented a device based on two concentric discs that simplified the use of Caesar ciphers. The substitution, i.e., the relative shift of the two alphabets, is determined by the relative rotation of the two disks (Figure 1.1).

Rumor has it that Alberti also considered changing the substitution within one message by turning the inner disc in his device. It is believed that this is how he discovered the so-called polyalphabetic ciphers, which are based on superpositions of Caesar ciphers with different shifts. For example, the first letter in the message can be shifted by 7, the second letter by 14, the third by 19, the fourth again by 7, the fifth by 14, the sixth by 19, and so on repeating the shifts 7, 14, 19 throughout the whole message. The sequence of numbers — in this example 7, 14, 19 — is usually referred to as a cryptographic key. Using this particular key we transform the message SELL into its concealed version, which reads ZSES.

As said, the message to be concealed is called the plaintext; the operation of disguising it is known as encryption. The encrypted plaintext is called the ciphertext or cryptogram. Our example illustrates the departure from a simple substitution; the repeated L in the plaintext SELL is enciphered differently in each case. Similarly, the two S's, in the ciphertext represent different letters in the plaintext: the first S corresponds to the letter E and the second to the letter L. This makes the straightforward frequency analysis of characters in ciphertexts obsolete. Indeed, polyalphabetic ciphers invented by the main contributors to the field at the time, such as Johannes Trithemius (1462–1516), Blaise de Vigenre (1523–1596), and Giovanni Battista Della Porta (1535–1615), were considered unbreakable for at least another 200 years. Indeed, Vigenre himself confidently dubbed his invention "le chiffre indéchiffrable" — the unbreakable cipher.

1.3   Not So Unbreakable

The first description of a systematic method of breaking polyalphabetic ciphers was published in 1863 by the Prussian colonel Friedrich Wilhelm Kasiski (1805–1881), but, according to some sources (for example, Simon Singh, *The Code Book*), Charles Babbage (1791–1871) had worked out the same method in private sometime in the 1850s.

The basic idea of breaking polyalphabetic ciphers is based on the observation that if we use N different substitutions in a periodic fashion then every Nth character in the cryptogram is enciphered with the same monoalphabetic cipher. In this case we have to find N, the length of the key and apply frequency analysis to subcryptograms composed of every Nth character of the cryptogram.

But how do we find N? We look for repeated sequences in the ciphertext. If a sequence of letters in the plaintext is repeated at a distance which is a multiple of N, then the corresponding ciphertext sequence is also repeated. For example, for N = 3, with the 7, 14, 19 shifts, we encipher TOBEORNOTTOBE as ACULCVUCMACUL:

T	O	B	E	O	R	N	O	T	T	O	B	E
A	C	U	L	C	V	U	C	M	A	C	U	L

The repeated sequence ACUL is a giveaway. The repetition appears at a distance 9; thus we can infer that possible values of N are 9 or 3 or 1. We can then apply frequency analysis to the whole cryptogram, to every third character and to every ninth character; one of them will reveal the plaintext. This trial and error approach becomes more difficult for large values of N, i.e., for very long keys.

In the 1920s, electromechanical technology transformed the original Alberti's disks into rotor machines in which an encrypting sequence with an extremely long period of substitutions could be generated, by rotating a sequence of rotors. Probably the most famous of them is the Enigma machine, patented by Arthur Scherbius in 1918.

A notable achievement of cryptanalysis was the breaking of the Enigma in 1933. In the winter of 1932, Marian Rejewski, a 27-year-old cryptanalyst working in the Cipher Bureau of the Polish Intelligence Service in Warsaw, mathematically determined the wiring of the Enigma's first rotor. From then on, Poland was able to read thousands of German messages encrypted by the Enigma machine. In July 1939 Poles passed the Enigma secret to French and British cryptanalysts. After Hitler invaded Poland and France, the effort of breaking Enigma ciphers continued at Bletchley Park in England. A large Victorian mansion in the center of the park (now a museum) housed the Government Code and Cypher School and was the scene of many spectacular advances in modern cryptanalysis.

1.4 Truly Unbreakable?

Despite its long history, cryptography only became part of mathematics and information theory in the late 1940s, mainly as a result of the work of Claude Shannon (1916–2001) of Bell Laboratories in New Jersey. Shannon showed that truly unbreakable ciphers do exist and, in fact, they had been known for over 30 years. They were devised in 1918 by an American Telephone and Telegraph engineer Gilbert Vernam and Major Joseph Mauborgne of the U.S. Army Signal Corps. They are called one-time pads or Vernam ciphers (Figure 1.2).

Both the original design of the one-time pad and the modern version of it are based on the binary alphabet. The plaintext is converted to a sequence of 0's and 1's, using some publicly known rule. The key is another sequence of 0's and 1's of the same length. Each bit of the plaintext is then combined with the respective bit of the key, according to the rules of addition in base 2:

$$0 + 0 = 0, \qquad 0 + 1 = 1 + 0 = 1, \qquad 1 + 1 = 0. \tag{1.2}$$

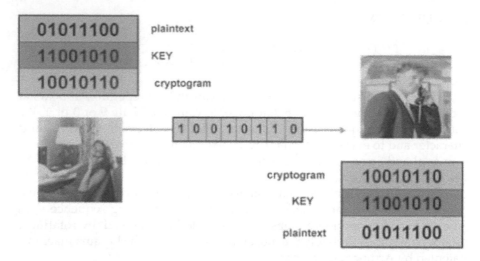

Figure 1.2   One-time pad.

The key is a random sequence of 0's and 1's, and therefore the resulting cryptogram, the plaintext plus the key, is also random and completely scrambled unless one knows the key. The plaintext can be recovered by adding (in base 2 again) the cryptogram and the key.

In the example above (shown in Figure 1.2), the sender, traditionally called Alice, adds each bit of the plaintext (01011100) to the corresponding bit of the key (11001010) obtaining the cryptogram (10010110), which is then transmitted to the receiver, traditionally called Bob. Both Alice and Bob must have exact copies of the key beforehand; Alice needs the key to encrypt the plaintext, Bob needs the key to recover the plaintext from the cryptogram. An eavesdropper, called Eve, who has intercepted the cryptogram and knows the general method of encryption but not the key, will not be able to infer anything useful about the original message. Indeed, Shannon proved that if the key is secret, the same length as the message, truly random, and never reused, then the one-time pad is unbreakable. Thus we do have unbreakable ciphers.

1.5   Key Distribution Problem

There is, however, a snag. All one-time pads suffer from a serious practical drawback, known as the key distribution problem. Potential users have to agree secretly and in advance on the key, a long, random sequence of 0's and 1's. Once they have done this, they can use the key for enciphering and deciphering, and the resulting cryptograms can be transmitted publicly, for example, broadcasted by radio, posted on the Internet, or printed in a newspaper, without compromising the security of the messages. But the key itself must be established between the sender and the receiver by means of a secure channel — for example, a secure telephone line, or via a private meeting or hand delivery by a trusted courier.

Such a secure channel is usually available only at certain times and under certain circumstances. So users far apart, in order to guarantee perfect security of subsequent cryptocommunication, have to carry around with them an enormous amount of secret and meaningless information (cryptographic keys), equal in volume to all the messages they might later wish to send. This is, to say the least, not very convenient.

Furthermore, even if a secure channel is available, this security can never be truly guaranteed. A fundamental problem remains because, in principle, any classical private channel can be monitored passively, without the sender or receiver knowing that the eavesdropping has taken place. This is because classical physics — the theory of ordinary-scale bodies and phenomena such as paper documents, magnetic tapes, and radio signals — allows all physical properties of an object to be measured without disturbing those properties. Since all information, including cryptographic keys, is encoded in measurable physical properties of some object or signal, classical theory leaves open the possibility of passive eavesdropping, because in principle it allows the eavesdropper to measure physical properties without disturbing them. This is not the case in quantum theory, which forms the basis for quantum cryptography. However, before we venture into quantum physics, let us mention in passing a beautiful mathematical approach to solving the key distribution problem.

The 1970s brought a clever mathematical discovery in the shape of "public-key" systems. The two main public-key cryptography techniques in use today are the Diffie–Hellman key exchange protocol [13] and the RSA encryption system (named after the three inventors, Ron Rivest, Adi Shamir, and Leonard Adleman) [24]. They were discovered in the academic community in 1976 and 1978, respectively. However, it was widely rumored that these techniques were known to British government agencies prior to these dates, although this was not officially confirmed until recently. In fact, the techniques were first discovered at the British Government Communication Headquarters in the early 1970s by James Ellis, who called them nonsecret encryption. In 1973, building on Ellis' idea, C. Cocks designed what we now call RSA, and in 1974 M. Williamson proposed what is essentially known today as the Diffie–Hellman key exchange protocol.

In the public-key systems, users do not need to agree on a secret key before they send the message. They work on the principle of a safe with two keys, one public key to lock it, and another private one to open it. Everyone has a key to lock the safe but only one person has a key that will open it again, so anyone can put a message in the safe but only one person can take it out. The systems avoid the key distribution problem but unfortunately their security depends on unproven mathematical assumptions. For example, RSA — probably the most popular public-key cryptosystem — derives its security from the difficulty of factoring large numbers. This means that if mathematicians or computer scientists come up with fast and clever procedures for factoring, the whole privacy and discretion of public-key cryptosystems could vanish overnight.

Indeed, we know that quantum computers can, at least in principle, efficiently factor large integers [19]. Thus in one sense public-key cryptosystems are already insecure: any RSA-encrypted message that is recorded today will become readable moments after the first quantum computer is switched on, and therefore RSA cannot be used for securely transmitting any information that will still need to be secret on that happy day. Admittedly, that day is probably decades away, but can anyone prove, or give any reliable assurance, that it is? Confidence in the slowness of technological progress is all that the security of the RSA system now rests on.

1.6   Local Realism and Eavesdropping

We shall now leave mathematics and enter the world of quantum physics. Physicists view key distribution as a physical process associated with sending information from one place to another. From this perspective, eavesdropping is a set of measurements performed on carriers of information. In order to avoid detection, an eavesdropper wants to learn about the value of a physical property that encodes information without disturbing it. Is such a passive measurement always possible?

In 1935, Albert Einstein together with Boris Podolsky and Nathan Rosen (EPR) published a paper in which they outlined how a "proper" fundamental theory of nature should look [15]. The EPR program required completeness ("In a complete theory there is an element corresponding to each element of reality.") and locality ("The real factual situation of the system A is independent of what is done with the system B, which is spatially separated from the former.") and defined the element of physical reality as "If, without in any way disturbing a system, we can predict with certainty the value of a physical quantity, then there exists an element of physical reality corresponding to this physical quantity." In other words, if we can know the value of some physical property without "touching" the system in any way, then the property must be physically real, i.e., it must have a determinate value, even before we measure it.

This world view is known as "local realism" and it implies possibilities of perfect eavesdropping. Indeed, this is exactly what the EPR definition of the element of reality means in the cryptographic context.

Einstein and his colleagues considered a thought experiment, on two entangled particles, that showed that quantum states cannot in all situations be complete descriptions of physical reality. The EPR argument, as subsequently modified by David Bohm [9], goes as follows. Imagine the singlet-spin state of two spin $\frac{1}{2}$ particles

$$| \Psi \rangle = \frac{1}{\sqrt{2}} (| \uparrow \rangle | \downarrow \rangle - | \downarrow \rangle | \uparrow \rangle) , \qquad (1.3)$$

where the single particle kets $| \uparrow \rangle$ and $| \downarrow \rangle$ denote spin up and spin down with respect to some chosen direction. This state is spherically symmetric, and the choice of direction does not matter. The two particles, which we label A and B, are emitted from a source and fly apart. After they are sufficiently separated so that they do not interact with each other, we can predict with certainty the x component of spin of particle A by measuring the x component of spin of particle B. Each measurement on B, in $\frac{1}{2}\hbar$ units, can yield two results, $+1$ (spin up) and -1 (spin down) and reveals the value of the x component of A. This is because the total spin of the two particles is zero, and the spin components of the two particles must have opposite values. The measurement performed on particle B does not disturb particle A (by locality) so the x component of spin is an element of reality according to the EPR criterion. By the same argument and by the spherical symmetry of state $|\Psi\rangle$ the y, z, or indeed any other spin components are also elements of reality. Therefore all the spin components must have predetermined values $+1$ or -1.

Local realism has experimental consequences. Consider two pairs of spin components, A_1 and A_2 pertaining to the particle A, and B_1 and B_2 pertaining to the particle B. A_1, A_2, B_1, and B_2 all have simultaneous definite values, either $+1$ or -1. Hence the quantity

$$Q = A_1(B_1 - B_2) + A_2(B_1 + B_2) \tag{1.4}$$

can have two different values, either -2 or $+2$, and consequently,

$$-2 \le \langle Q \rangle \le 2, \tag{1.5}$$

where $\langle Q \rangle$ stands for the average value of Q. This inequality is known as the Bell inequality [3] or more precisely as the CHSH inequality [11].

Both quantum-mechanical predictions and experiments show that for two particles in the singlet state, $\langle AB \rangle = -\vec{a} \cdot \vec{b}$, $\vec{a}$ and $\vec{b}$ are the unit vectors specifying the directions of the spin components of particles A and B, respectively. This leads to a violation of the CHSH inequality (Equation (1.5)). For if we choose $\vec{a}_i$ and $\vec{b}_j$ in the x–y plane, perpendicular to the trajectory of the particles emitted from the source, and characterized by the azimuthal angles $\phi_1^a = 0$, $\phi_2^a = \frac{1}{2}\pi$, and $\phi_1^b = \frac{1}{4}\pi$, $\phi_2^b = \frac{3}{4}\pi$ then $\langle Q \rangle = -2\sqrt{2}$. Local realism is refuted, which opens possibilities of constructing key distribution schemes that will always detect eavesdropping.

Please note that any theory that refutes local realism, be it quantum or post-quantum, opens such possibilities. Even if quantum mechanics is refuted sometime in the future and a new physical theory is conjectured, as long as the new theory refutes local realism, possibilities for post-quantum cryptography are wide open.

1.7   Quantum Key Distribution

1.7.1   Entanglement-Based Protocols

Let us take advantage of the CHSH inequality within the quantum theory. The key distribution is performed via a quantum channel that consists of a source that emits pairs of spin $\frac{1}{2}$ particles in the singlet state as in

Equation (1.3). The particles fly apart along the z-axis toward the two legitimate users of the channel, Alice and Bob, who, after the particles have separated, perform measurements and register spin components along one of three directions, given by unit vectors $\vec{a}_i$ and $\vec{b}_j$ (i, $j = 1, 2, 3$), respectively, for Alice and Bob. For simplicity, both $\vec{a}_i$ and $\vec{b}_j$ vectors lie in the x–y plane, perpendicular to the trajectory of the particles, and are characterized by azimuthal angles: $\phi_1^a = 0$, $\phi_2^a = \frac{1}{4}\pi$, $\phi_3^a = \frac{1}{2}\pi$ and $\phi_1^b = \frac{1}{4}\pi$, $\phi_2^b = \frac{1}{2}\pi$, $\phi_3^b = \frac{3}{4}\pi$. Superscripts a and b refer to Alice's and Bob's analyzers, respectively, and the angle is measured from the vertical x-axis. The users choose the orientation of the analyzers randomly and independently for each pair of incoming particles. Each measurement can yield two results, $+1$ (spin up) and -1 (spin down) and can reveal one bit of information.

After the transmission has taken place, Alice and Bob can announce in public the orientations of the analyzers they have chosen for each particular measurement and divide the measurements into two separate groups: a first group for which they used different orientations of the analyzers and a second group for which they used the same orientation of the analyzers. They discard all measurements in which either or both of them failed to register a particle at all. Subsequently Alice and Bob can reveal publicly the results they obtained, but within the first group of measurements only. This allows them to establish the value of $\langle Q \rangle$, which if the particles were not directly or indirectly "disturbed" should be very close to $-2\sqrt{2}$. This assures the legitimate users that the results they obtained within the second group of measurements are anticorrelated and can be converted into a secret string of bits — the key.

An eavesdropper, Eve, cannot elicit any information from the particles while in transit from the source to the legitimate users, simply because there is no information encoded there. The information "comes into being" only after the legitimate users perform measurements and communicate in public afterwards. Eve may try to substitute her own prepared data for Alice and Bob to misguide them, but as she does not know which orientation of the analyzers will be chosen for a given pair of particles, there is no good strategy to escape being detected. In this case her intervention will be equivalent to introducing elements of *physical reality* to the spin components and will lower $\langle Q \rangle$ below its "quantum" value.

1.7.2   Prepare and Measure Protocols

Instead of tuning into an external source of entangled particles, Alice and Bob may also rely on the Heisenberg uncertainty principle. Suppose a spin $\frac{1}{2}$ particle is prepared in one of the four states, say spin up and down along the vertical x-axis ($|\uparrow\rangle$, $|\downarrow\rangle$) and spin up and down along the horizontal y-axis ($|\rightarrow\rangle$, $|\leftarrow\rangle$). Then the two x states $|\uparrow\rangle$ and $|\downarrow\rangle$ can be distinguished by one measurement and the two y states $|\rightarrow\rangle$ and $|\leftarrow\rangle$ by another measurement. The measurement that can distinguish between the two x states will give a completely random outcome, when applied to distinguish between the two y

states and vice versa. If, for each incoming particle, the receiver performing the measurement is not told in advance which type of spin (x or y) was prepared by the sender, then the receiver is completely lost and unable to determine the spin value. This can be used for the key distribution.

Alice and Bob agree on the bit encoding, e.g., $|\uparrow\rangle = 0 = |\rightarrow\rangle$, $|\downarrow\rangle = 1 = |\leftarrow\rangle$, and Alice repeatedly prepares one of the four quantum states, choosing randomly out of $|\uparrow\rangle$, $|\downarrow\rangle$, $|\rightarrow\rangle$, and $|\leftarrow\rangle$. She then sends it to Bob, who randomly chooses to measure either the x or the y spin component. After completing all the measurements, Alice and Bob discuss their data in public so that anybody can listen, including their adversary Eve. Bob tells Alice which spin component he measured for each incoming particle and she tells him "what should have been measured." Alice does not disclose which particular state she prepared, and Bob does not reveal the outcome of the measurement, so the actual values of bits are still secret. Alice and Bob then discard those results in which Bob failed to detect a particle and those for which he made measurements of the wrong type. They then compare a large subset of the remaining data. Provided no eavesdropping has taken place, the result should be a shared secret that can be interpreted by both Alice and Bob as a binary key.

But let us suppose there is an eavesdropper, Eve. Eve does not know in advance which state will be chosen by Alice to encode a given bit. If she measures this bit and resends it to Bob, this may create errors in Bob's readings. Therefore in order to complete the key distribution Alice and Bob have to test their data for discrepancies. They compare in public some randomly selected readings and estimate the error rate; if they find many discrepancies, they have reason to suspect eavesdropping and should start the whole key distribution from scratch. If the error rate is negligibly small, they know that the data not disclosed in the public comparison form a secret key. No matter how complex and subtle is the advanced technology and computing power available to the eavesdropper, the "quantum noise" caused inevitably by each act of tapping will expose each attempt to gain even partial information about the key.

1.8 Security Proofs

Admittedly the key distribution procedures described above are somewhat idealized. The problem is that there is in principle no way of distinguishing noise due to an eavesdropper from innocent noise due to spurious interactions with the environment, some of which are presumably always present. All good quantum key distribution protocols must be operable in the presence of noise that may or may not result from eavesdropping. The protocols must specify for which values of measurable parameters Alice and Bob can establish a secret key and provide a physically implementable procedure that generates such a key. The design of the procedure must take into account that an eavesdropper may have access to unlimited quantum computing power.

The best way to analyze eavesdropping in the system is to adopt the entanglement-based protocol and the scenario that is most favorable for eavesdropping, namely that Eve herself is allowed to prepare and deliver all the pairs that Alice and Bob will subsequently use to establish a key. This way we take the most conservative view, which attributes all disturbance in the channel to eavesdropping, even though most of it (if not all) may be due to innocent environmental noise. This approach also applies to the prepare and measure protocols because they can be viewed as special cases of entanglement-based protocols, e.g., the source of entangled particles can be given either to Alice or to Bob. It is prudent to assume that Eve has disproportional technological advantage over Alice and Bob. She may have access to unlimited computational power, including quantum computers; she may monitor all the public communication between Alice and Bob in which they reveal their measurement choices and exchange further information in order to correct errors in their shared key and to amplify its privacy. In contrast, Alice and Bob can only perform measurements on individual qubits and communicate classically over a public channel. They do not have quantum computers, or any sophisticated quantum technology, apart from the ability to establish a transmission over a quantum channel.

The search for good security criteria under such stringent conditions led to early studies of quantum eavesdropping [17,28] and finally to the first proof of the security of key distribution [12]. The original proof showed that the entanglement-based key distributions are indeed secure and noise-tolerant against an adversary with unlimited computing power as long as Alice and Bob can implement quantum privacy amplification. In principle, quantum privacy amplification allows us to establish a secure key over any distance, using entanglement swapping [29] in a chain of quantum repeaters [2,14]. However, this procedure, which distills pure entangled states from corrupted mixed states of two qubits, requires a small-scale quantum computation. Subsequent proofs by Inamori [21] and Ben-Or [4] showed that Alice and Bob can also distill a secret key from partially entangled particles using only classical error correction and classical privacy amplification [6,7].

Quantum privacy amplification was also used by Lo and Chau to prove the security of the prepare and measure protocols over an arbitrary distance [22]. A concurrent proof by Mayers showed that the protocol can be secure without Alice and Bob having to rely on the use of quantum computers [23]. The same conclusion, but using different techniques, was subsequently reached by Biham et al. [8]. Although the two proofs did not require quantum privacy amplification, they were rather complex. A nice fusion of quantum privacy amplification and error correction was proposed by Shor and Preskill, who formulated a relatively simple proof of the security of the BB84 [5] protocol based on virtual quantum error correction [25]. They showed that a protocol that employs quantum error-correcting code to prevent Eve from becoming entangled with qubits that are used to generate the key reduces to the BB84 augmented by classical error correction and classical privacy amplification. This proof has been further extended by Gottesman and Lo [20] for two-way public communication to allow for a higher bit error rate in BB84 and by Tamaki et al. [26] to proof the security of the B92 protocol.

More recently, another simple proof of the BB84, which employs results from quantum communication complexity, has been provided by Ben-Or [4] and a general proof based on bounds on the performance of quantum memories has been proposed by Christandl et al. [30].

Let us also mention in passing that apart from the scenario that favors Eve, i.e., Eve has access to quantum computers while Alice and Bob do not, there are interesting connections regarding the criteria for the key distillation in commensurate cases, i.e., when Alice, Bob, and Eve have access to the same technology, be it classical or quantum [18,10,1].

1.9 Concluding Remarks

Quantum cryptography was discovered independently in the U.S. and Europe. The first one to propose it was Stephen Wiesner, then at Columbia University in New York, who, in the early 1970s introduced the concept of quantum conjugate coding [27]. He showed how to store or transmit two messages by encoding them in two "conjugate observables" such as linear and circular polarization of light, so that either, but not both, of which may be received and decoded. He illustrated his idea with a design of unforgeable bank notes. A decade later, building upon this work, Charles H. Bennett of the IBM T. J. Watson Research Center and Gilles Brassard of the Université de Montreal, proposed a method for secure communication based on Wiesner's conjugate observables [5]. However, these ideas remained by and large unknown to physicists and crytologists. In 1990, independently and initially unaware of the earlier work, the current author, then a Ph.D. student at the University of Oxford, discovered and developed a different approach to quantum cryptography based on peculiar quantum correlations known as quantum entanglement [16]. Since then, quantum cryptography has evolved into a thriving experimental area and is quickly becoming a commercial proposition.

This brief overview has only scratched the surface of the many activities that are presently being pursued under the heading of quantum cryptography. It is focused solely on the development of theoretical concepts that led to creating unbreakable quantum ciphers. The experimental developments, although equally fascinating, are left to the other contributors to this book. I have also omitted many interesting topics in quantum cryptography that go beyond the key distribution problem. Let me stop here hoping that even the simplest outline of quantum key distribution has enough interesting physics to keep you entertained for a while.

References

1. A. Acin, N. Gisin, and V. Scarani, Security bounds in quantum cryptography using d-level systems, *Quant. Inf. Comp.*, 3(6), 563–580, November 2003.
2. H. Aschauer and H.-J. Briegel, A security proof for quantum cryptography based entirely on entanglement purification, *Phys. Rev. A*, 66, 032302, 2002.
3. J.S. Bell, *Physics*, 1, 195, 1964.

4. M. Ben-Or, Simple security proof for quantum key distribution. On-line presentation available at http://www.msri.org/publications/ln/msri/2002/qip/ben-or/1/index.html.

5. C.H. Bennett and G. Brassard, Quantum cryptography, public key distribution and coin tossing, in *Proceedings of International Conference on Computer Systems and Signal Processing*, IEEE, 1984, p. 175.

6. C.H. Bennett, G. Brassard, C. Crépeau, and U. Maurer, Generalized privacy amplification, *IEEE Trans. Inf. Theory*, 41(6), 1915–1923, 1995.

7. C.H. Bennett, G. Brassard, and J.-M. Robert, Privacy amplification by public discusion. *SIAM Journal on Computing*, 17(2), 210–229, 1988.

8. E. Biham, M. Boyer, P.O. Boykin, T. Mor, and V. Roychowdhury, A proof of the security of quantum key distribution, in *Proceedings of the Thirty-Second Annual ACM Symposium on Theory of Computing (STOC)*, ACM Press, New York, 2000, p. 715. quant-ph/9912053.

9. D. Bohm, *Quantum Theory*, New York: Prentice Hall, 1951.

10. D. Bruss, M. Christandl, A. Ekert, B.-G. Englert, D. Kaszlikowski, and C. Macchiavello, Tomographic quantum cryptography: equivalence of quantum and classical key distillation, *Phys. Rev. Lett.*, 91, 097901, 2003. quant-ph/0303184.

11. J.F. Clauser, M.A. Horne, A. Shimony, and R.A. Holt, *Phys. Rev. Lett.*, 23, 880, 1969.

12. D. Deutsch, A. Ekert, R. Jozsa, C. Macchiavello, S. Popescu, and A. Sanpera, Quantum privacy amplification and the security of quantum cryptography over noisy channels, *Phys. Rev. Lett.*, 77, 2818–2821, 1996. Erratum, ibid., 80, 2022–2022, 1998, quant-ph/9604039.

13. W. Diffie and M.E Hellman, *IEEE Trans. Inf. Theory*, IT-22, 644, 1976.

14. W. Dür, H.-J. Briegel, J.I. Cirac, and P. Zoller, Quantum repeaters based on entanglement purification, *Phys. Rev. A*, 59, 169–181, 1999.

15. A. Einstein, B. Podolsky, and N. Rosen, Can quantum-mechanical description of physical reality be considered complete? *Phys. Rev.*, 47, 777, 1935.

16. A. K. Ekert, Quantum cryptography based on Bell's theorem, *Phys. Rev. Lett.*, 67(6), 661, 1991.

17. A.K. Ekert and B. Huttner, Eavesdropping techniques in quantum cryptosystems, *Journal of Modern Optics*, 41, 2455–2466, 1994. Special issue on Quantum Communication.

18. N. Gisin and S. Wolf, in *Advances in Cryptology — CRYPTO'00*, Lecture Notes in Computer Science, Springer–Verlag, 2000, pp. 482–500.

19. S. Goldwasser, ed., *Proceedings of the 35th Annual Symposium on the Foundations of Computer Science*, IEEE Computer Society Press, Los Alamitos, CA, 1994.

20. D. Gottesmann and H.-K. Lo., Proof of security of quantum key distribution with two-way classical communication, *IEEE Trans. Inf. Theory*, 49(2), 457–475, 2003, quant-ph/0105121.

21. H. Inamori, Security of EPR-based quantum key distribution, quant-ph/0008064.

22. H.-K. Lo and H. F. Chau, Unconditional security of quantum key distribution over arbitrarily long distances, *Science*, 283(5410), 2050–2056, 1999.

23. D. Mayers, Unconditional security in quantum cryptography, quant-ph/9802025, 1998.

24. R. Rivest, A. Shamir, and L. Adleman, On digital signatures and public-key cryptosystems, Technical Report MIT/LCS/TR-212, MIT Laboratory for Computer Science, January 1979.

25. P. W. Shor and J. Preskill, Simple proof of security of the bb84 quantum key distribution protocol, *Phys. Rev. Lett.*, 85(2), 441–444, 2000; quant-ph/0003004.
26. K. Tamaki, M. Koashi, and N. Imoto, Unconditionally secure key distribution based on two nonorthogonal states, *Phys. Rev. Lett.*, 90, 167904, 2003.
27. S. Wiesner, Conjugate coding, *Sigact News*, 15(1), 78–88, 1983; Originally written c. 1970 but them unpublished.
28. A.C.-C. Yao, Security of quantum protocols against coherent measurements, in *Proceedings of the 27th ACM Symposium on the Theory of Computing*, ACM Press, 1995, pp. 67–75.
29. M. Zukowski, A. Zeilinger, M. Horne, and A.K. Ekert, Event-ready detectors, Bell experiment via entanglement swapping, *Phys. Rev. Lett.*, 71, 4287–4290, 1993.
30. M. Christandl, R. Renner, and A. Ekert, A generic security proof for quantum key distribution, quant-ph/0402131.

chapter 2

Quantum Communications with Optical Fibers

N. Gisin, S. Iblisdir, W. Tittel, and H. Zbinden
University of Geneva

Contents

2.1   A Geneva-Biased Introduction 18
2.2   Time-Bin Qubits and Higher Dimensions 19
2.3   Faint Laser Quantum Cryptography: The Plug & Play
       Configuration 23
       2.3.1   Basics of Faint Laser Quantum Key Distribution 23
       2.3.2   A Practical Realization: The Plug & Play
               Configuration 24
2.4   Two-Photon Quantum Cryptography 27
       2.4.1   Single-Photon Based Realizations 27
       2.4.2   Entanglement-Based Realizations 27
               2.4.2.1   Long-Distance Quantum Correlation 29
               2.4.2.2   Quantum Key Distribution 32
2.5   The Future of Quantum Cryptography 34
       2.5.1   Quantum Cryptography and Entanglement 35
       2.5.2   PNS Attacks and Countermeasures 36
       2.5.3   Three- and Four-Photon Quantum
               Communication 36
       2.5.4   Conclusions and Outlook 39
Acknowledgments 39
References 40

Abstract

This chapter reviews experimental and theoretical achievements of the Group of Applied Physics (GAP) at University of Geneva in the domain of quantum communication. All work presented can be motivated by the goal to render experimental quantum key distribution simple and robust, and to devise means to extend the maximum transmission distance in spite of technical imperfections like lack of single-photon sources, lossy quantum channels and non-perfect detectors. In detail, we present an auto-aligning "plug & play" system for quantum key distribution based on faint laser pulses, two entanglement-based systems, teleportation in quantum relay configuration and finally entanglement swapping. All experiments take advantage of photons at telecommunication wavelengths and optical fibers, and use time-bin encoding which enables us to demonstrate the different protocols over distances of a few kilometers to several tens of kilometers. In addition, we developed a new protocol for quantum key distribution which also enables extending the maximum transmission distance in spite of so-called photon number splitting eavesdropper attacks and non-ideal faint laser pulses instead of true single photons.

2.1   A Geneva-Biased Introduction

Quantum communication is the natural follow-up of classical communication into the era of quantum technologies. Accordingly, its natural wavelengths are the same as those used in today's fiber optics communication. However, most groups opted for shorter wavelengths, because they originate from academic research in quantum optics, which traditionally uses visible or near-infrared light. Additionally, until recently, detectors sensitive to single photons existed only at those shorter wavelengths, roughly below 1µm. In contrast to most groups, in Geneva we originate from a research group dealing with classical telecom physics. Hence, when we started our activities in quantum communication, we decided to go for the telecom wavelengths (see Figure 2.1). This implied from the very beginning the development of single-photon detectors at these (at the time) exotic wavelengths. The next step was to choose the appropriate degree of freedom to encode quantum information. We opted for a solution that we believe is better adapted to optical fibers though at first sight less obvious than polarization: the time-bins (see Section 2.2).

Our group is also known for having introduced Faraday mirrors into the field of quantum communication. It should be stressed that such mirrors have been invented by Professor M. Martinelli from Milan [1]. In Geneva, we first used them for a fiber-based sensor [2], a development that turned out to be much less successful than our activity on quantum cryptography. Actually, the main outcome of this sensor project has been the idea of using Faraday mirrors in quantum communication [82]. The nonreciprocal Faraday effect is also used in many other crucial components in quantum optics experiments: isolators and circulators are based on this effect.

In the following pages, we review first the concept of time-bins, in Section 2.2. Next, pseudo-single photon quantum cryptography is presented in Section 2.3, followed by two-photon quantum cryptography (which includes tests of Bell inequalities as a natural child), in Section 2.4. Finally, in Section 2.5.3 we briefly comment on three- and four-photon applications: quantum teleportation, entanglement swapping and quantum relays.

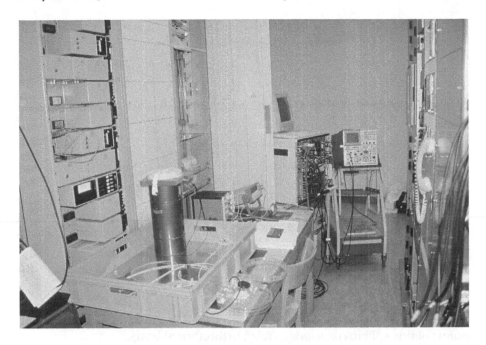

Figure 2.1 First quantum cryptography experiment outside the lab, in 1995, at the Swisscom telecommunication center. (From A. Muller et al., *Europhys. Lett.* 33(5), 335–339, 1996.)

Let us emphasize that more specific reviews exist, for instance and among others, on time-bit qubits and related experiments in [6], on quantum cryptography in [7], on quantum cryptography and entanglement in [8], and on Bell inequalities and useful entanglement in [9].

2.2 Time-Bin Qubits and Higher Dimensions

The fundamental constituent underlying most quantum communication protocols is the qubit. As a quantum analogue of a classical bit, a qubit is simply a two-level quantum system. The quantum information carried by a qubit is its state, which is, according to quantum mechanics, described by a normalized vector in $\mathbb{C}^2$. If we let $\{|0\rangle, |1\rangle\}$ denote an orthonormal basis of $\mathbb{C}^2$, the state of a qubit reads $a_0|0\rangle + a_1|1\rangle$. Also, quantum mechanics prescribes that vectors that are identical up to a global phase factor essentially describe the same state. Thus the general state of a qubit can be written as

$$|\psi\rangle = \cos\frac{\theta}{2}|0\rangle + \sin\frac{\theta}{2}e^{i\phi}|1\rangle, \tag{2.1}$$

where $\phi \in [0:2\pi]$, and $\theta \in [0:\pi]$. A nice feature of this parametrization is that it allows us to represent conveniently all (pure) qubit states on the surface of a

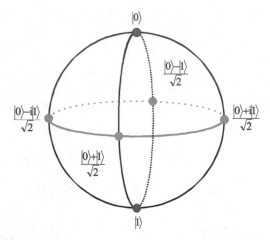

Figure 2.2 The Poincaré sphere representation of qubit states.

sphere, the so-called Poincaré sphere (see Figure 2.2). Any couple of antipodal points of this sphere corresponds to an orthonormal basis.

There are many physical systems whose degrees of freedom can implement a qubit. Quantum dots [10] or electronic levels of atoms in cavities QED [11] are examples of such systems. For quantum communication, though, the most natural candidate is provided by photons, because they are relatively immune to decoherence when compared to other implementations, and because we generally do not need to make them interact much. Polarization is often used to encode photonic qubits. Here we will describe an encoding that is particularly well-suited for our purposes: the so-called time-bin qubits [6].

To prepare a time-bin qubit, one processes a short one-photon pulse at one input port of an unbalanced interferometer [83] (see the left part of Figure 2.3). One then gets a superposition of a state where one photon is in the lower arm of the interferometer with a state where one photon is in the upper arm of the interferometer. If the path length difference between the two arms chosen is much larger than the spread of the pulse, it is possible to use a switch to process on the same fiber the part of the pulse corresponding to the short path and the part corresponding to the long path. Now $|0\rangle$ represents a one-photon state localized in the late time-bin and $|1\rangle$ a one-photon state localized in the early time-bin, as depicted in the middle of Figure 2.3. So, we see that by tuning the phase and the coupling of the (left) interferometer, we can prepare any (pure) qubit state.

As shown in the right part of Figure 2.3, we can use a similar interferometer to perform measurements on a qubit. With the help of the switch, the first time-bin is sent to the long arm and the second time-bin to the short one. The path length difference between the two arms of this interferometer is chosen to annihilate the time difference between the two time-bins, and make them arrive simultaneously at the variable coupler. A photon is then detected by

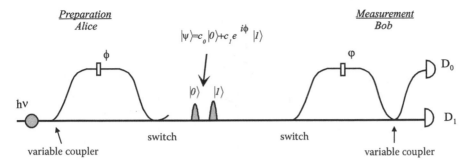

Figure 2.3 Setup used to prepare and analyze time-bin qubit states.

one of two detectors, D_0 or D_1. Upon tuning the coupling and the phase shift of the right interferometer of Figure 2.3, we thus can perform any simple qubit measurement.

We now turn to the issue of generalizing the above scheme and consider qudits, that is, d-dimensional quantum systems. In the context of quantum information processing, these higher-dimensional systems are interesting to consider: in quantum key distribution, for example, such systems carry intrinsically more information than qubits and are more resilient to noise [12]. Another example is nonlocality tests: it has been possible to construct Bell inequalities for qudits for which the required detector efficiencies are significantly lower than for qubits [13].

It is trivial to extend the two schemes we have described to the case of qudits. One just adds as many arms to the interferometer as the dimension of the system one wants to prepare. Again, a variable coupler distributes the impinging one-photon state among the d arms, with a weight amplitude c_j for the jth arm, resulting in a state that is a superposition of states with one photon in each arm. The path length difference between any two arms is much larger than the pulse spread. On each arm j, a phase shift ϕ_j is applied. Finally, a switch is used to direct the d parts of this state onto the same fiber. The state we get reads $\Sigma_{j=1}^{d} c_j e^{i\phi_j} |j\rangle$, where $|j\rangle$ denotes a state of one photon lying in the time-bin j. First results along the above lines can be found in [14–16]. Note that all projective Von Neumann measurements can, in principle, be implemented using this approach.

Let us now see how entangled states can be produced using a nonlinear crystal in which spontaneous parametric down conversion (SPDC) occurs. In SPDC, the impinging photon, called the pump photon, is (probabilistically) converted into two photons called signal and idler. Thus if the pump photon is in a superposition state of two time-bins (Equation (2.1)), the two-photon state emerging from the crystal will be

$$|\Psi\rangle = \cos\frac{\theta}{2}|0, 0\rangle + e^{i\phi}\sin\frac{\theta}{2}|1, 1\rangle, \tag{2.2}$$

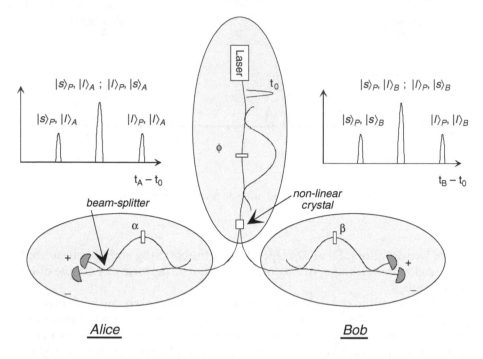

Figure 2.4   Setup used to prepare and analyze time-bin entangled qubit states.

where $|0, 0\rangle$ denotes a state in which two photons (one signal photon and one idler photon) are in the time-bin 0, and $|1, 1\rangle$ denotes a state where two photons are in the time-bin 1.

These (bipartite) entangled states can be analyzed using the setup shown in Figure 2.4. Again, if the two analyzing interferometers have the same path length difference as the interferometer located before the crystal, it is possible (using adequate phase shifters and couplers) to measure each qubit on any basis and thus detect entanglement and/or nonlocality. Considering maximally entangled states, for example $(\cos(\theta/2) = 1/\sqrt{2})$, the coincidence count rate between Alice and Bob detectors as a function of the phases ϕ, α, β is given by $R_c(\phi, \alpha, \beta) = 1 + \cos(\alpha + \beta - \phi)$. These rates $R_c(\phi, \alpha, \beta)$ cannot be explained by any local hidden variable theory [17] and are a typical signature of the nonlocal character of the state (Equation (2.2)). Finally, we note that, again, the scheme depicted in Figure 2.4 generalizes to the case of qudits; adding more arms to the interferometer before the crystal allows the creation of entangled time-bin qudit states. And these states can be analyzed by Alice and Bob using switches, passive optical elements, and photodetectors.

To conclude this section, let us (briefly) comment on experimental issues regarding the schemes we have described. For simplicity's sake, we will restrict ourselves to qubits and consider only the Figures 2.3 and 2.4. Our first comment concerns the source. We do not have (to date) single photon sources useful for long distance communications. Therefore, one rather uses a source that emits short pulses in a weak coherent state $|\phi\rangle = |\phi_0\rangle + |\phi_1\rangle + |\phi_2\rangle$, where $|\phi_0\rangle, |\phi_1\rangle, |\phi_2\rangle$ denote the zero-photon, one-photon and two-photon contribution to ϕ respectively. If the source is weak enough, the two-photon contribution can be neglected. Then, most of the time, no detector clicks. But if one detector has clicked [84], the process which has occurred is exactly the one that would have happened, had we had a genuine single-photon source. Our second comment is about the use of optical switches. In practice, such devices are quite lossy. Therefore, one can replace them with passive couplers. There are then three possible arrival times for the photon, according to the possible combinations of paths through the two interferometers. If eventually one photon is detected in the central time window, such variable couplers behave as ideal optical switches. To summarize these two comments, we can say that an ideal scheme can be replaced with a nonideal one at the price of postselection. In addition, note that the detection of a photon in the left or right satellite peak corresponds to a projection onto the early ($|1\rangle$) or late ($|0\rangle$) time-bin, respectively. The use of a passive coupler in the analyzer does not therefore engender additional losses, but guarantees a passive and random selection of the measurement basis.

2.3 Faint Laser Quantum Cryptography: The Plug & Play Configuration

2.3.1 Basics of Faint Laser Quantum Key Distribution

Quantum communication is about sending qubits from Alice to Bob. Quantum cryptography (or better, quantum key distribution (QKD)) is about establishing a secret key (a string of random, secret bits) between Alice and Bob [7]. It exploits a fundamental principle of quantum mechanics, which is that one cannot completely determine an unknown quantum state without disturbing it [85]. The idea is that Alice and Bob can see whether the exchanged key has been eavesdropped by checking to see if it has been disturbed. For this purpose the qubits, in practice photons, must be sent and measured according to a suitable protocol, the most widely known and applied being the BB84 protocol [18]. In this protocol, Alice sends, for example, photons with four different linear polarization states, of 0 or 90 and 45 or 135 degrees, respectively. These are the four states indicated on the equator of the Poincaré sphere of Figure 2.2, the states on the poles representing right and left circularly polarized light, respectively (note that from the discussion held in the previous section, Alice could as well use time-bins; see the next section). Bob cannot distinguish unambiguously between the four different states, however, he is measuring randomly along one of the two measurement bases (0/90 or 45/145 degrees) and gets a conclusive result in half of the cases.

An eavesdropper Eve cannot, exactly like Bob, obtain every time a conclusive result, as the basis used is unknown. Therefore a simple intercept-and-resend strategy, in which Eve measures the incoming photon and sends a new one to Bob (prepared according to the measurement result), will fail. Indeed, half the time Eve will use a noncompatible basis, and the reemitted photon will introduce errors in the key of Alice and Bob. Hence, by checking the errors in the key, Alice and Bob are able to reveal the presence of an eavesdropper. This fact can be shown to hold for any eavesdropping strategy, as perfectly elaborated as it might be; see [7] and references therein. Only after the key exchange do Alice and Bob tell each other which basis they used for each photon. They can establish a sifted key by attributing to each polarization state a bit value (0 or 1), keeping events with compatible bases and discarding the others. They correct the errors that were introduced by the imperfect key exchange and apply a procedure called privacy amplification, which allows them to eliminate the information that Eve might have acquired (supposing her action is at the origin of the detected error). It can be shown that Alice and Bob can distill a random, perfectly secret key as long as the error rate is smaller than 11% (respectively 15%, depending on the assumptions one wants to make on a potential eavesdropper) [7]. However, the efficiency of the distillation becomes very small for error rates above 10%.

In the original papers introducing the idea, quantum cryptography is naturally based on single photons. However, true single photon sources are very difficult to realize experimentally. In fact, only a few QKD experiments with single photons have been reported up to now [19,20]. In addition, presently available sources are not yet suitable for (fiber) QKD — their wavelengths are not adapted to fiber telecommunications. Therefore, today's QKD setups, which have been successfully tested and are ready to be commercialized, are based on faint laser pulses. In addition, a couple of rather proof-of-principle experiments using photon-pairs have been performed (see Section 2.4).

Faint laser pulses are light pulses with a Poisson distributed number of photons and an average number of photons well below 1. Their creation is very simple but there are two disadvantages. First, since most faint laser pulses contain in fact zero photons, the bit rate and the signal-to-noise ratio are considerably reduced. Second, since some faint laser pulses contain two or more photons, there is some opportunity for eavesdropping, see [73], the discussion held in Section 2.5.2, and Chapter 6 of this book. However, it can be shown that faint pulse QKD can be mathematically secure over distances up to 100 km with present technology [86]. In particular, we recently presented a new protocol [21,22] that can increase the range and bit rate of secure faint laser QKD.

2.3.2   A Practical Realization: The Plug & Play Configuration

The principles of QKD we have explained use polarized photons. The first experiment through 30 cm [23] of air used several states of polarization as

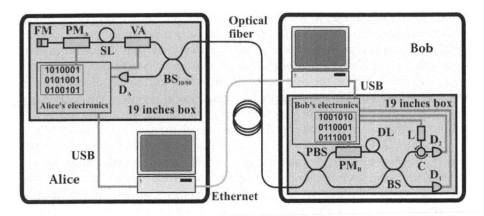

Figure 2.5 Self-aligned plug & play system (L: laser diode, APD: avalanche photodiode, BS: beam splitter, C: circulator, PM_j: phase modulator, PBS: polarizing beam splitter, DL: optical delay line, FM: Faraday mirror, D: classical detector).

well as the latest free space experiments over more than 20 km [24]. For long distance QKD through optical fibers, however, polarization is probably not the appropriate choice. Polarization is not maintained in optical fibers [87]. Therefore, polarization-based QKD through a long optical fiber link asks for a permanent compensation for the rapidly changing evolution of the polarization state in the fiber. In contrast, as seen in Section 2.2, any arbitrary time-bin qubit can be created and measured using interferometers. Indeed, one of the first QKD experiments in optical fibers used essentially the setup shown in Figure 2.3 [25,88]. In such a setup, one difficulty is to stabilize the arm length difference in both interferometers down to a fraction of a wavelength during the whole key exchange. However, by folding the interferometer of Figure 2.3 and sending light pulses back and forth, it is possible to design an auto-compensating interferometer [4,26]. Since no continuous alignment is required this design is also called plug & play configuration.

Let us have a closer look at the plug & play setup (see Figure 2.5) [27]. A strong laser pulse (@1550 nm) emitted at Bob's (L) is separated at a first 50/50 beam splitter (BS). The two pulses impinge on the input ports of a polarization beam splitter (PBS) after having traveled through a short arm and a long arm, including a phase modulator (PM_B) and a 50 ns delay line (DL), respectively. All fibers and optical elements at Bob's are polarization maintaining. The linear polarization is turned by 90 degrees in the short arm, so the two pulses exit Bob's setup by the same port of the PBS. The pulses travel down to Alice, are reflected on a Faraday mirror (FM), are attenuated well below an average photon number per pulse of 1 (VA), and come back orthogonally polarized [89]. In turn, both pulses now take the other path at Bob's and arrive at the same time at the BS, where they interfere. Since the two pulses take the same path, inside Bob's apparatus in reversed order, this interferometer is autocompensated. Finally, the photon is detected either in D1 or after passing through the circulator (C) in D2. To implement the BB84 protocol, Alice's electronics detects the incoming pulse and applies randomly a phase shift of 0, π, π or, $3\pi/2$ on the second pulse with a phase modulator PM_A. Bob's electronics generates the laser pulses, and chooses the measurement basis upon applying a 0 or $\pi/2$ phase shift on the first pulse on its way back. Finally, Bob gates the detector when the photon is supposed to arrive and registers the click. Alice and Bob's systems are connected to computers that communicate with each other via ethernet or any other public classical channel.

The stability of the autoaligning interferometer has been tested in the field [27]. Over terrestrial cables longer than 60 km and aerial cables longer than 10 km, interference visibilities higher than 99.5% have been measured. This means that the interference is almost perfect and does not introduce significant bit errors. Indeed, almost all bit errors (or the quantum bit error rate (QBER)) are due to the noise of the detectors. On one hand, the probability that a photon arrives at Bob's decreases exponentially with the distance due to the losses in the optical fiber; on the other hand, the dark count probability is constant (a dark count is an event where a single-photon detector clicks although there is no photon). Therefore the QBER increases exponentially with distance, which limits the range of QKD, knowing that key distillation becomes very inefficient above a QBER of 10%. At present, InGaAs avalanche photodiode are used as photon counters for the 1550 nm telecom window. These detectors have dark count probabilities of the order of 10^{-6} per gate (time window of about 1 ns), limiting QKD to distances of about 100 km. A couple of experiments have been performed over distances in this range [27–29]. Today, pulse rates are in the order of a few MHz, leading to net key creation rates in the order of 1500 and 50 Hz over 20 and 70 km, respectively [27]. Again, the performance of the available photon counters will eventually limit the maximal pulse rate. Nevertheless, faint laser QKD has left the lab: commercial systems are available (Figure 2.6) [30] and it is also being considered seriously from a military perspective, see Chapter 3 of this book.

Figure 2.6   Commercial QKD-system based on the plug & play setup.

2.4 Two-Photon Quantum Cryptography

As we have seen in Section 2.2, time-bin qubits realized using faint pulses can be employed to encode and to transmit quantum information. The success of the plug & play scheme (see Section 2.3) proves them to be particularly well suited for long distance QKD via optical fibers. However, like any experimental implementation, faint-pulse-based QKD schemes do not perfectly meet the theoretical ideal for QKD. Nevertheless, this does not necessarily render it unsecure. For instance, the approximation of single photons by faint pulses only limits the maximum transmission span. First, in connection with detector noise, the nonvanishing vacuum component leads to a higher QBER compared to true single photons and thus to smaller maximum distance (see the discussion at the end of Section 2.3.2). Second, the possibility of photon number splitting attacks, based on the existence of pulses containing more than one photon, places an upper bound on the tolerable transmission losses, hence on the distance (see Section 2.5.2). Finally, even if a specific realization of QKD works well for a particular setting, it might not be adapted to different working conditions. For instance, it is difficult to generate true random numbers and to implement the corresponding settings in a high-speed system. In this section, we will show how photon-pair-based QKD systems, although more complicated to implement, might help to overcome some limits of faint-pulse-based systems. A further expansion of some of the ideas can be found in Section 2.5.3, where we will elaborate on quantum teleportation as a quantum relay.

2.4.1 Single Photon Based Realizations

In order to get around one drawback of faint pulses, i.e., the high probability of having zero photons in a pulse, a good idea is to replace the faint pulse source by a SPDC-based photon-pair source (see Figure 2.7b) where one photon serves as a trigger to indicate the presence of the other [31,62]. In this case, Alice can remove the vacuum component of her source, and Bob's detectors are only activated whenever she sends at least one photon. In principle, it is thus possible to achieve a probability of emitting a nonempty pulse equal to one [90]. This leads to a higher sifted key rate (assuming the same trigger rate as in the faint-pulse case) and a lower QBER for a given distance (for given losses) and therefore to a larger maximum span. It is important to note that the use of photon pairs created by SPDC does not avoid problems with multiphoton pulses. For a given mean number, the probabilities that a nonempty pulse contains more than one photon, or pair, respectively, are essentially the same [32]. Therefore, the possibility of multiphoton splitting eavesdropping attacks exists as well.

2.4.2 Entanglement-Based Realizations

The potential of a source creating photon pairs is not restricted to creation of two photons at the same time — one serving as a trigger for the other one.

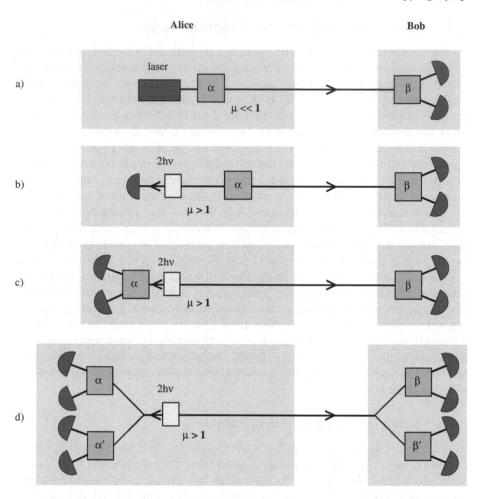

Figure 2.7 Single-photon based quantum cryptography using (a) a faint-pulse source, (b) a two-photon source, (c) entanglement-based quantum cryptography with active, and (d) with passive choice of bases. 2hν denotes the photon-pair source, μ the mean number of photons per pulse, and the parameters α and β the settings of the qubit-creation and analyzing devices, respectively.

It is possible to use the full quantum correlation of *entangled* pairs to generate identical keys at Alice's and Bob's and to test the presence of an eavesdropper via a test of a Bell inequality (see Figure 2.7c). This beautiful application of tests of Bell inequalities has been pointed out by Ekert in 1991 [33] — without knowing about the "discovery" of quantum cryptography by Bennett and Brassard seven years earlier. The setup is similar to the one used to test Bell inequalities, with the exception that Alice and Bob each have to choose from three and not only two different bases. Depending on the basis chosen for each specific photon pair, the measured data is used to establish the sifted key (whenever the choice of bases enables perfect correlation), or to test a Bell inequality, or it is discarded.

The security of the Ekert protocol is easy to understand: the action of the eavesdropper Eve acquiring knowledge about the state of the photon traveling to Bob can be described as adding probabilistic hidden variables (hidden in the sense that only the eavesdropper knows about their value). If she gets full information about all states, i.e., if the whole set of photons analyzed by Bob can be described by hidden variables, a Bell inequality cannot be violated anymore. If Eve has only partial knowledge, the violation is less than maximal, and if no information has leaked out at all, Alice and Bob observe a maximal violation.

Despite its beauty, the Ekert protocol is not very efficient concerning the ratio of transmitted bits to the sifted key length. As pointed out in 1992 by Bennett et al. [34] as well as by Ekert et al. [35], protocols originally devised for single-photon schemes can also be used for entanglement-based realizations. This is not surprising if one considers Alice's action as a nonlocal state preparation for the photon traveling to Bob. Interestingly, it turns out that, if the perturbation of the quantum channel (the QBER) assuming the BB84 protocol is such that the Alice–Bob mutual Shannon information equals Eve's maximum Shannon information, then the Clauser–Horne–Shimony–Holt (CHSH) Bell inequality [36] cannot be violated any more [37,38]. Although this seems very natural in this case, it is not clear yet to what extent the connection between the security of quantum cryptography and the violation of a Bell inequality can be generalized.

In the following, we will first briefly present a test of Bell inequalities over a distance of 10 km and then comment on some experimental realizations of quantum key distribution based on photon-pair correlation. We refer the reader to Chapter 3 of this book or a survey of other experiments of quantum communication with entangled photons.

2.4.2.1 Long-Distance Quantum Correlation
As mentioned before, a requirement for entanglement-based QKD is the generation and the transmission of entangled two-photon states with a degree of correlation that enables them to violate the CHSH Bell inequality. The largest spatial separation to date — a distance of 10 km — has been achieved in tests that were carried out in our group in 1997 and 1998 [5,39,41] (see Figure 2.8). Further developments in 2002 and 2003 led to the observation of quantum correlation between analyzers connected with up to 50 km of fiber [42–44], but, in opposition to the first-mentioned experiments, this was realized with fiber on a spool. All setups relied on photon pairs at telecommunication wavelengths (either 1310 nm or 1550 nm) suitable for transmission in standard telecommunication optical fibers (for a photo of a compact telecommunication wavelength photon-pair source see Figure 2.9), and fiber-optical interferometers equipped with Faraday mirrors to compensate polarization effects.

The first series of experiments [5,39,41] took advantage of energy-time entanglement created by spontaneous parametric down-conversion. This type

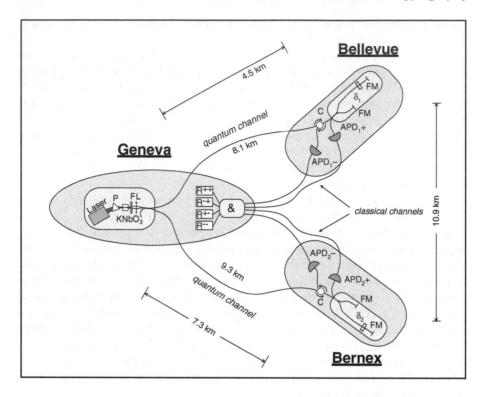

Figure 2.8 Experimental arrangement for a test of Bell inequalities with measurements made more than 10 km apart. Source (in Geneva) and observer stations (interferometers in Bellevue and Bernex, respectively) were connected using the Swisscom fiber optic telecommunications network.

of entanglement can be seen as the "continuous" version of time-bin entanglement and does not belong to the class of entangled qubits: signal and idler photons are again emitted simultaneously within their coherence time, but the emission time of the pair must be this time described as a continuous distribution of infinitely many possibilities (not only two) bounded only by the large coherence time of the pump laser. Energy-time entanglement has been proposed in the context of a test of Bell inequalities in 1989 by Franson [45] (10 years before time-bin entanglement [46]) and demonstrated for the first time in 1992 and 1993 by Brendel et al. [47] and Kwiat et al. [48], respectively. Despite its formal difference, the measurement of energy-time entangled photons is similar to the one introduced for time-bin entanglement (Section 2.2): the correlated photons are separated, and sent to equally unbalanced interferometers and the coincidence count rates of photons emerging from the interferometer simultaneously (either both photons having passed via the short, or both via the long arm) are measured. Therefore, the detection process determines the two time-bins that were superposed in the local qubit projection measurements, thereby collapsing the continuously entangled state to

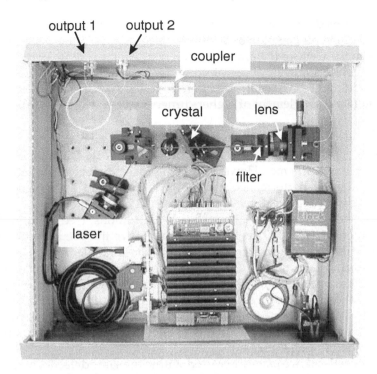

Figure 2.9 Photo of our entangled photon pair source used in the first long-distance field test of quantum correlation. Note that the whole source including temperature and power control of the diode laser fits into a box of only $40 \times 45 \times 15$ cm^3.

the familiar entangled time-bin qubit state (Equation (2.2)). As long as the coherence time of the pump laser is large compared to the travel-time difference Δt introduced in the interferometer, i.e., as long as pair emission with time difference Δt is coherent, the coincidence count rate shows a sinusoidal dependence on the sum of the phase in both interferometers.

After having realized in 1999 that it is possible to render "continuous" energy-time entanglement "discrete" by replacing the high coherent pump by a succession of a finite number of short, i.e., well localized pump pulses [46], we performed a couple of experiments based on time-bin entangled photons. For instance, in 2002, we demonstrated the robustness of maximally as well as partially entangled qubits over 11 km of fiber on a spool [42]. Finally, in 2003, we could extend the separation of the analyzing interferometers to more than 50 km, again taking advantage of fiber on a spool [44]. These investigations showed that energy-time as well as time-bin entanglement is well suited for QKD via optical fibers over long distances.

To conclude this section, let us briefly mention further directions of research that emerged from the developments addressed above. First, the possibility of distributing entanglement over long distances enabled it to falsify relativistic nonlocality (or multisimultaneity) [49–51] and made it possible to put an improved lower bound on the speed of "quantum information," i.e., the speed of the propagation of a hypothetical collapse of the wave function: analysis in the Geneva (laboratory) reference frame led to $2/3 \times 10^7$ c (c: the speed of light) [50], while observation in the frame of the cosmic microwave background radiation fixed the bound to $1.5 \cdot 10^4 c$ [91,52]. Second, we started to investigate entanglement of higher-dimensional systems [14–16].

2.4.2.2 Quantum Key Distribution

Obviously, entanglement-based QKD is more complicated to implement than faint-pulse based schemes. However, as long as technological limitations like those on detector performance and the lack of efficient true single-photon sources remain, it also features advantages.

First, as with single-photon based realizations (see Section 2.4.1), entanglement-based QKD enables Alice to remove the vacuum component of the generated pulses sent to Bob. Actually, the entanglement-based case is even more efficient, since the optical losses in Alice's preparation device also are now eliminated, as can be seen from Figure 2.7c. In addition, depending on the position of the source, the probability of detecting a photon at Bob's, conditioned on detection of its twin at Alice's, can be further increased. This probability is optimal if the source is located in the middle, resulting in a minimal quantum-bit error rate.

Second, even if two pairs are created within the same detection window — hence two photons travel towards Bob within the same pulse — they are completely independent and do not carry the same qubit, although they are prepared in states belonging to the same basis [92]. Only the photon forming a pair with the photon detected at Alice's is in a definite quantum state; the other photon is in a completely mixed state. Therefore, eavesdropping attacks based on multiphoton pulses do not apply in entanglement-based QKD. However, multiphoton pulses lead to errors at Bob's, who detects from time to time a photon that is not correlated to Alice's [53].

The third advantage is directly linked to the one mentioned before: beyond the passive state preparation, it is even possible to achieve a passive choice of bases using a setup similar to the one depicted in Figure 2.7d: no external switch that forces all photons in a pulse to be prepared or measured in a given basis is required, but each photon independently "chooses" its basis and bit value. Therefore, no fast random number generator or active change of basis is required.

Finally, the fact that the possibility of distillation of a secret key is intrinsically linked to the possibility of violating the CHSH Bell inequality ensures that the different states are not distinguishable through other uncontrolled degrees of freedom [54]. For instance, assuming a faint-pulse or single-photon based scheme where all states are generated by a different setup [24], differences in the wavelength of the photons encoding nonorthogonal states would enable the eavesdropper to acquire full knowledge about the quantum state sent without perturbing it: in frequency space, all states would be orthogonal.

Using entangled states, such a correlation would prevent from violating a Bell inequality and could thus be discovered easily.

Although all Bell experiments intrinsically contain the possibility for entanglement-based QKD, only a few experiments have been devised in order to allow a fast change of measurement bases. Interestingly enough, the first experiment that fulfills this criterion is the test of Bell inequalities using time-varying analyzers, performed by Aspect et al. in 1982 [55] — at a time when quantum cryptography was still unknown, even the single-photon based version. More experiments enabling active [56] and passive [41] change of bases followed in 1998 and 1999. However, as with the first-mentioned experiment, the bases chosen for the measurements are chosen in order to allow a test of Bell inequalities and not to establish a secret key. The first experiments that allowed the distribution of a quantum key were finally performed in 2000 [57–59], and more followed in 2001 [60], 2002 [61] and 2004 [43,44].

All our experiments [43,44,59,60] incorporated a passive choice of bases. In 2000 and again in 2003, we took advantage of an original solution offered by time-bin entanglement [44,59]. As mentioned before, only detections in the central time window correspond to a projection on a basis with eigenstates represented on the equator on the qubit sphere. Interestingly, detections in one of the two remaining windows are not unwanted events that have to be discarded after postselection but correspond to projection onto the $|0\rangle$ or $|1\rangle$ states. Hence, time-bin entanglement offers passive choice of bases for free. While the first experiment [59] has been performed over a short distance, we could recently extend this distance to 50 km (fiber on a spool) with actively stabilized interferometers, which is the longest transition span for entanglement-based QKD to date [44].

In 2001, we realized a QKD system based on energy-time entanglement [60], similar to the test of Bell inequalities mentioned in Section 2.4.2.1. However, this realization took advantage of an asymmetric setup, with the source close to Alice's, instead of a setup designed for tests of Bell inequalities with the source located roughly in the middle between Alice and Bob. This makes it possible to employ high efficiency and low noise silicon avalanche photodiodes detectors (that cannot detect photons at telecommunication wavelengths) at Alice's side, together with photons at nontelecommunication wavelength (that cannot be sent through long fibers due to enhanced absorption). The second photon (again at 1550 nm telecommunication wavelength) is transmitted through 8.5 km of dispersion shifted fiber to a fiber optical interferometer, equipped, as usual, with Faraday mirrors. The passive choice of bases was implemented using polarization multiplexing (see Figure 2.10), similar to Figure 2.7d. We recently repeated this experiment using 30 km of standard fiber, managing the dispersion by filtering or compensation [43].

Finally, in addition to the mentioned two-party QKD schemes, we reported in 2001 a proof-of-principle demonstration of quantum secret sharing (three-party quantum cryptography) in a laboratory experiment [63]; see also Chapter 7 of this book. This rather new protocol enables Alice to send key

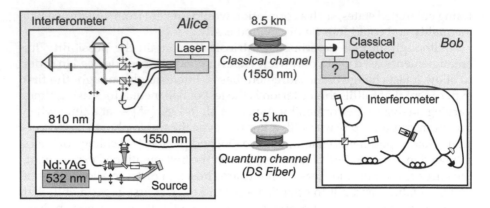

Figure 2.10 Schematics of the energy-time entanglement-based quantum QKD system.

material to Bob and Charlie in a way that neither Bob nor Charlie alone have any information about Alice's key. However, when comparing their data, they have full information. The goal of this protocol is to force them to collaborate.

In contrast to implementations using three-particle GHZ states [64,65], pairs of time-bin entangled qubits were used to mimic the necessary quantum correlation of three entangled qubits, albeit only two photons exist at the same time (see Figure 2.11). This is possible thanks to the symmetry between the preparation interferometer acting on the pump pulse and the interferometers analyzing the down-converted photons. Indeed, the data describing the emission of a bright pump pulse at Alice's is equivalent to the data characterizing the detection of a photon at Bob's and Charlie's: all specify a phase value and an output or input port, respectively. Therefore, the emission of a pump pulse can be considered as a detection of a photon with 100% efficiency, and the scheme features a much higher coincidence count rate compared to the initially proposed GHZ-state type schemes.

2.5   The Future of Quantum Cryptography

What is the future of quantum cryptography? It is likely that it is still in its infancy. Conceptually, as alluded to above, it is deeply rooted in the basic aspect of quantum physics, i.e., in entanglement. On the applied side, it is already attracting attention (and money) from industries and investors [30]. On the experimental side, quite a lot of progress is still to be expected, mainly on two major specifications: the secret bit rate and the distance. These are not independent but may be addressed in different, complementary, ways. First, there is a mere technological approach: improve the detectors and/or the

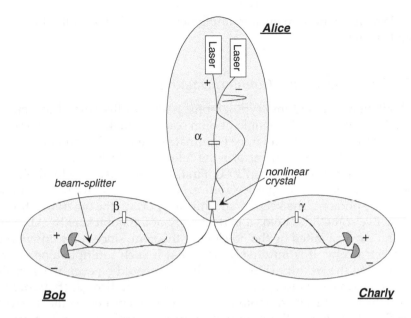

Figure 2.11   Basic setup for three-party quantum secret sharing using pseudo-GHZ states.

source. Second, develop new protocols, like SARG (see [21,22] and the next paragraph) and continuous variables [66]. Finally, exploit the ideas of quantum teleportation and entanglement swapping to develop quantum relays [67,68] and quantum repeaters [69]. Below, we briefly comment on our contributions to the above-mentioned ideas to extend quantum cryptography.

2.5.1   Quantum Cryptography and Entanglement

Quantum cryptography is very much related to quantum physics. In particular *quantum entanglement* is conceptually closely related to *secret classical information*.

For example, consider a situation where three parties, A, B, E, share many copies of a (pure) state ψ_{ABE}, where the partial state $\rho_{AB} = \text{Tr}_E \psi_{ABE}$ is a two-qubit entangled state. In this situation, it has been proven that ρ_{AB} can be mapped by single-copy measurements into a probability distribution $P(a,b)$ containing secret correlations between A and B, i.e., were A and B to produce the distribution $P(a,b)$ using only classical means, they would need to use secret bits as an extra resource [71]. Conversely, if Alice and Bob share a quantum state ρ_{AB} from which a probability distribution $P(a,b)$ containing secret correlations can be obtained, then ρ_{AB} is entangled. These facts illustrate the central role played by quantum cryptography in quantum information. Also on the experimental side, the central role played by entanglement can be seen even in implementations that actually do not use entanglement [8].

2.5.2 PNS Attacks and Countermeasures

One limitation to quantum cryptography is due to the small but nonzero probability that a (pseudo-) single-photon source actually emits two or more photons. If this is the case and if the legitimate partners Alice and Bob use the BB84 protocol, then Eve can perform the following attack, known as photon number splitting (PNS) attack [73,74]. First, Eve measures the total photon-number in each pulse leaving Alice's office. Next, whenever the pulse contains two or more photons, she keeps one, while teleporting the other(s) to Bob. In this way Eve sometimes holds a perfect copy of the qubit sent by Alice and is thus no longer limited by the no-cloning theorem, since Alice unwillingly offered her a copy. It is intuitively clear that if such multiphotons are too frequent, then the security is lost. The situation could be even worse if Eve can block the single-photon pulses and Bob does not notice this (Bob could be fooled if the missing single-photon pulses are compensated by the increased probability that he gets the multiphoton pulses thanks to the — assumed perfect — teleportation). Hardware-based countermeasures involve a strong reference pulse [76] or modulating pulse intensity [78]. Recently we imagined a way to attenuate the effect of PNS attacks, with a mere change of the sifting part of the BB84 protocol. The basic idea is to encode bits into nonorthogonal states. In this way, even if Eve holds a perfect copy, she cannot extract full information about the encoded bit. The same holds true, of course, for Bob. But Bob is in a much more favorable situation than Eve: he may perform an unambiguous state discrimination [75,77] and simply declare to Alice whether he was successful or not [21,22].

2.5.3 Three- and Four-Photon Quantum Communication

Finally, let us mention how quantum teleportation and entanglement swapping could be exploited to extend the distances over which quantum cryptography is secure. The first remark is that detectors are noisy: there is a finite dark count probability, which is clearly independent of the distance. Secondly, the signal, on the contrary, i.e., the rate at which Bob detects photons, decreases exponentially with the distance. Hence, for any detector/channel pair, there is a distance limit beyond which quantum cryptography is unpractical [7]. Attractive ways around this limitation are quantum repeaters and quantum relays. The idea of a relay consists in dividing the channel into n equal trunks [67,68]. Halves of the inner nodes contain a two-photon source (EPR source of entangled photons as described in Section 2.4); the other halves contain a (partial) Bell measurement [79]; see Figure 2.12. In this way, Bob's detector is activated only if all the Bell measurements were

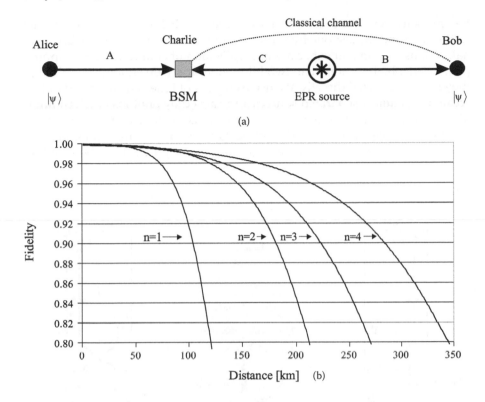

Figure 2.12 (a) Quantum teleportation as a quantum relay. (b) Fidelity of the transmitted quantum state as a function of the distance for different configurations. Direct transmission ($n = 1$), with an EPR source in the middle ($n = 2$), teleportation ($n = 3$), and entanglement swapping ($n = 4$). We assume that the fidelity is only affected by the detector's noise. The curves are plotted for a realistic dark count probability $D = 10^{-4}$ per ns and a fiber attenuation of 0.25 db/km.

successful. Consequently, the bad chance of a dark count is reduced to the cases where a photon is lost in the last trunk only. This intuitive idea can be elaborated (see [68]) though one should keep in mind that this trick does not improve the bit rate (quite the opposite actually; the poor efficiency of the Bell measurement reduces the bit rate). But it provides a good motivation to work on quantum teleportation. Furthermore, once quantum memories exist, combining them with the quantum relays will provide a working quantum repeater that will extend quantum cryptography to unlimited distances and bit rates.

Recently we demonstrated quantum teleportation in optical fibers over a significant distance. In a first experiment [80], the receiver was set in a lab 55 meters away and connected to the EPR source by 2 km of fiber on a spool. In a second [81] experiment we extended this to a case of 3 × 2 km: three trunks each of 2 km of fiber, the first one between Alice's source and the

Bell measurement, the second trunk between the Bell measurement and the EPR source and the last one from there to Bob's receiver (see Figure 2.12). Finally, let us mention an experimental realization of entanglement swapping [70], where photons more than two kilometers apart are entangled, although they never interacted directly. We use two pairs of time-bin entangled qubits created in spatially separated sources and carried by photons at telecommunication wavelengths. A partial Bell state measurement is performed with one photon from each pair, which projects the two remaining photons, formerly independent, onto an entangled state (see Figure 2.13). The visibility

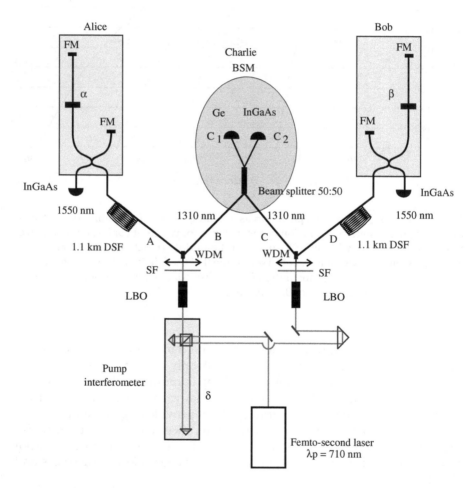

Figure 2.13 Experimental setup for entanglement swapping. The pump laser is a mode-locked femtosecond Ti-sapphire laser producing 200 fs pulses at a wavelength of 710 nm with a repetition rate of 75 MHz. After the crystals, the pump beams are blocked with silicon filters (SF). The Faraday mirrors (FM) are used to compensate polarization fluctuations in the fiber interferometers.

obtained after the swapping process was high enough to infer a violation of Bell inequalities between the remaining photons.

2.5.4 Conclusions and Outlook

To summarize this review, we presented experimental and theoretical achievements of GAP Optique in the domains of fundamental research and quantum communication. In contrast to work done in other groups, our experiments take advantage of photons at telecommunication wavelengths, standard optical fibers and time-bin qubits, enabling implementations over distances between a few kilometers and several tens of kilometers, depending on the application. All work is closely connected to quantum key distribution and can be summarized under the common aspect of increasing the performance of quantum key distribution in spite of the lack of true single photon sources, of lossy quantum channels and imperfect detectors. Starting with an auto aligning plug & play system for quantum cryptography based on faint laser pulses, we turned to experiments with pairs of time-bin entangled photons that render eavesdropping attacks based on photon number splitting ineffective, and that make it possible (in principle) to increase the maximum transmission span as engendered by the probabilistic arrival of photons at Bob's receiver station. Extending our work to three photons and pursuing the idea to reduce the effect of loss in the quantum channel, we then demonstrated quantum teleportation in a quantum relay configuration. Finally, adding a fourth photon, we could recently demonstrate the fundamental concept of entanglement swapping (or teleportation of entanglement), again in the spirit of a quantum relay. In parallel to these experimental issues, we proposed a new protocol that also makes it possible to attenuate the effect of photon number splitting attacks, and we pointed out a general link between quantum entanglement and classical information.

Quantum key distribution is now commercialized by different companies. In the short term, improvements should mainly be related to detectors: this could considerably increase the actual key creation rates and sightly increase the range. However, if we want to be able to implement QKD on scales of several hundreds of kilometers, we need quantum repeaters. Therefore, mid- and long-term research efforts are focused on long-distance entanglement swapping and quantum memories.

The future of quantum communication appears bright with still plenty of opportunities for both experimental and theoretical physicists addicted to conceptual issues and for those who are more engineering oriented.

Acknowledgments

Financial support by the Swiss OFES within the European projects RESQ and RAMBOQ and by the Swiss NCCR *Quantum Photonics* is acknowledged. We thank all collaborators from GAP Optique for their contributions to the various experimental and theoretical investigations mentioned in this review.

References

1. M. Martinelli, *Opt. Commun.*, 72, 341–344, 1989; M. Martinelli, *J. Modern Opt.*, 39, 451–455, 1992.
2. J. Breguet, N. Gisin and J.P. Pellaux, *Sensors and Actuators*, A48, 29–35, 1995. J. Breguet and N. Gisin, *Optics Lett.*, 20, 1447–1449, 1985.
3. A. Muller, H. Zbinden and N. Gisin, *Europhys. Lett.*, 33 (5), 335–339, 1996. See also *Nature* 378, 449, 1995.
4. A. Muller, T. Herzog, B. Huttner, W. Tittel, H. Zbinden and N. Gisin, *Applied Phys. Lett.*, 70, 793–795, 1997.
5. W. Tittel, J. Brendel, H. Zbinden, and N. Gisin, *Phys. Rev. Lett.*, 81, 3563–3566, 1998; H. Zbinden, N. Gisin, J. Brendle, and W. Tittel, *Phys. Rev. A*, 63, 022111/1–10, 2001.
6. W. Tittel and G. Weihs, *Quantum Information and Computation*, 1, 3–56, 2001.
7. N. Gisin, G. Ribordy, W. Tittel and H. Zbinden, *Rev. Modern Phys.*, 74, 145–195, 2002.
8. N. Gisin and N. Brunner, quant-ph/0312011, to appear in the proceedings of the *Les Houches* summer school, 2003.
9. A. Acin, N. Gisin, L. Masanes, and V. Scarani, *Int. J. Quant. Inf.*, 2, 23, 2004.
10. G. Burkard, H.-A. Engel, and D. Loss, *Fortschritte der Physik*, 48, 965, 2000.
11. J.M. Raimond, M. Brune and S. Haroche, *Rev. Mod. Phys.*, 73, 565, 2001.
12. N. Cerf, M. Bourennane, A. Karlsson and N. Gisin, *Phys. Rev. Lett.*, 88, 127902, 2002; D. Bruss and C. Macchiavello, *Phys. Rev. Lett.*, 88, 127901, 2002; M. Bourennane, A. Karlsson, G. Björn, N. Gisin and N. Cerf, *J. Phys. A: Math. and Gen.*, 35, 10065, 2002; D. Kaszlikowski, A. Gopinathan, Y. C. Liang, L. C. Kwek, B.-G. Englert, *Phys. Rev.*, A 70, 032306, 2004.
13. S. Massar, *Phys. Rev.*, A 65, 032121, 2002.
14. H. de Riedmatten, I. Marcikic, H. Zbindin and Nicolas Gisin, *Quantum Information and Computation*, 2, 42, 2002; quant-ph/0204165.
15. R. T. Thew, A. Acín, H. Zbinden and N. Gisin, *Quantum Information and Computation*, in print; quant-ph/0307122.
16. H. de Riedmatten, I. Marcikic, V. Scarani, W. Tittel, H. Zbinden and N. Gisin, *Phys. Rev.*, A 69, 050304, 2004.
17. J. S. Bell, *Speakable and Unspeakable in Quantum Mechanics*, Cambridge, U.K.: Cambridge University Press, 1987.
18. C.H. Bennett and G. Brassard, *Proc. Int. Conf. Computers, Systems & Signal Processing*, Bangalore, India, 1984, 175–179.
19. E. Waks, K. Inoue, C. Santori, D. Fattal, J. Vuckovic, G. S. Solomon and Y. Yamamoto, *Nature*, 420, 762, 2002.
20. A. Beveratos, R. Brouri, T. Gacoin, A. Villing, J.-P. Poizat and P. Grangier, *Phys. Rev. Lett.*, 89, 187901, 2002.
21. A. Acin, V. Scarani and N. Gisin, *Phys. Rev.*, A 69, 012309 2004; quant-ph/0302037.
22. V. Scarani, A. Acin, G. Ribordy and N. Gisin, *Phys. Rev. Lett.*, 92, 057901, 2004.
23. C.H. Bennett, F. Bessette, G. Brassard, L. Salvail, and J. Smolin, *Experimental Quantum Cryptography J. Cryptology*, 5, 3–28, 1992.
24. Ch. Kurtsiefer et al., *Nature*, 419, 450, 2002.
25. Ch. Marand and P.D. Townsend, *Optics Lett.*, 20, 1695, 1995.

26. D. Bethune and W. Risk, *IEEE J. Quantum Electron.*, 36, 340–347, 2000; M. Bourennane, F. Gibson, A. Karlsson, A. Hening, P. Jonsson, T. Tsegaye, D. Ljunggren and E. Sundberg, *Opt. Express*, 4, 383–387, 1999.
27. D. Stucki, N. Gisin, O. Guinnard, G. Ribordy and H. Zbinden, *New Journal of Physics*, 4, 41.1–41.8, 2002.
28. H. Kosaka et al., *Electr. Lett.*, 39, 1199–1201, 2003.
29. Y. Zhiliang, Ch. Gobby and A.J. Shields, preprint 2003.
30. www.idQuantique.com, www.MagiQtech.com.
31. C.K. Hong and L. Mandel, *Phys. Rev. Lett.*, 56, 58, 1986.
32. H. de Riedmatten, V. Scarani, I. Marcikic, A. Acin, W. Tittel, H. Zbinden, and N. Gisin, *J. Mod. Opt.*, 51, 1637, 2004.
33. A.K. Ekert, *Phys. Rev. Lett.*, 67, 661–663, 1991.
34. C. Bennett, G, Brassard, and N.D. Mermin, *Phys. Rev. Lett.*, 68, 557, 1992.
35. A. Ekert, J.G. Rarity, P.R. Tapster, and G.M. Palma, *Phys. Rev. Lett.*, 69, 1293, 1992.
36. J.F. Clauser, M.A. Horne, A. Shimony, and R.A. Holt, *Phys. Rev. Lett.*, 23, 880, 1969.
37. N. Gisin and B. Huttner, *Phys. Lett.*, A 228, 13, 1997.
38. C.A. Fuchs et al., *Phys. Rev.*, A 56, 1163, 1997.
39. W. Tittel, J. Brendel, B. Gisin, T. Herzog, H. Zbinden, and N. Gisin, *Phys. Rev.*, A 57, 3229, 1998.
40. W. Tittel, J. Brendel, H. Zbinden, and N. Gisin, *Phys. Rev. Lett.*, 81, 3563, 1998.
41. W. Tittel, J. Brendel, N. Gisin, and H. Zbinden, *Phys. Rev.*, A 59, 4150, 1999.
42. R. T. Thew, S. Tanzilli, W. Tittel, H. Zbinden, N. Gisin, *Phys. Rev.*, A 66, 062304-1/5, 2002.
43. S. Fasel, N. Ginsin, G. Ribordy, and H. Zbinden, *Eur. Phys. J. D*, 30, 143, 2004.
44. I. Marcikic et al., arXiv quant-ph/0404124.
45. J.D. Franson, *Phys. Rev. Lett.*, 62, 2205–2208, 1989.
46. J. Brendel, N. Gisin, W. Tittel and H. Zbinden, *Phys. Rev. Lett.*, 82, 2594, 1999.
47. J. Brendel, E. Mohler, and W. Martienssen, *Europhys. Lett.*, 20, 575, 1992.
48. P.G. Kwiat, A.M. Steinberg, and R.Y. Chiao, *Phys. Rev.*, A 47, 2472, 1993.
49. A. Suarez, and V. Scarani, *Phys. Lett.*, A 232, 9, 1997.
50. H. Zbinden, J. Brendel, N. Gisin, and W. Tittel, *Phys. Rev.*, A 63, 022111, 2001.
51. A. Stefanov, H. Zbinden, N. Gisin and A. Suares, *Phys. Rev. Lett*, 88, 120404, 2002.
52. V. Scarani, W. Tittel, H. Zbinden, and N. Gisin, *Phys. Lett.*, A 276, 1, 2000.
53. I. Marcikic, H. de Riedmatten, W. Tittel, V. Scarani, H. Zbinden, and N. Gisin, *Phys. Rev.*, A 66, 062308, 2002.
54. D. Mayers and A. Yao, *Proceedings of the 39th IEEE Conference on Foundations of Computer Science*, Los Alamitos, California, 1998, p. 503.
55. A. Aspect, J. Dalibard, and G. Rogers, *Phys. Rev. Lett.*, 49, 1804, 1982.
56. G. Weihs, T. Jennewein, C. Simon, H. Weinfurter, and A. Zeilinger, *Phys. Rev. Lett.*, 81, 5039, 1998.
57. T. Jennewein, C. Simon, G. Weihs, H. Weinfurter, and A. Zeilinger, *Phys. Rev. Lett.*, 84, 4729–4732, 2000.
58. D. Naik, C. Peterson, A. White, A. Bergiund, and P. Kwiat, *Phys. Rev. Lett.*, 84, 4733–4736, 2000.
59. W. Tittel, J. Brendel, H. Zbinden, and N. Gisin, *Phys. Rev. Lett.*, 84, 4737, 2000.
60. G. Ribordy, J. Brendel, J.D. Gautier, N. Gisin, and H. Zbinden, *Phys. Rev.*, A 63, 012309, 2001.

61. D.G. Enzer, P.G. Hadley, R. J. Hughes, C.G. Peterson, and P.G. Kwiat, *New J. Phys.*, 4, 45.1, 2002.
62. S. Fasel, O. Alibart, A. Beveratos, S. Tanzilli, H. Zbinden, P. Baldi and N. Gisin, arXiv report quant-ph/0408136.
63. W. Tittel, H. Zbinden, and N. Gisin, *Phys. Rev.*, A 63, 042301, 2001.
64. M. Zukowski, A. Zeilinger, M. Horne, and H. Weinfurter, *Acta Phys. Pol.*, A 93, 187–195, 1998.
65. M. Hillery, V. Buzek, and A. Berthiaume, *Phys. Rev.*, A 59, 1829–1834, 1999.
66. Grosshans F. et al., *Nature*, 421, 238–240, 2003.
67. B.C. Jacobs, T.B. Pittman, and J.D. Franson, *Phys. Rev.*, A 66, 052307, 2002.
68. D. Collins, N. Gisin and H. de Riedmatten, quant-ph/0311101
69. H. Briegel, W. Dür, J. I. Cirac and P. Zoller, *Phys. Rev. Lett.*, 81, 5932, 1998.
70. H. de Riedmatten, I. Marcikic, J.A.W. van Houwelingen, W. Tittel, H. Zbinden, H. Gisin, LANL arXiv quant-ph/0409093.
71. A. Acin and N. Gisin, quant-ph/0310054.
72. T. Acin, N. Gisin and L. Massanes, *Phys. Rev. Lett.*, 91, 167901, 2003.
73. N. Lütkenhaus, *Phys. Rev.*, A 61, 052304, 2000; G. Brassard, N. Lütkenhaus, T. Mor and B.C. Sanders, *Phys. Rev. Lett.*, 85, 1330, 2000.
74. G. Brassard, N. Lütkenhaus, T. Mor and B.C. Sanders, *Phys. Rev. Lett.*, 85, 1330–1333, 2000.
75. A. Peres, *Quantum Theory: Concepts and Methods*, (Kluwer, Dordrecht, 1998), section 9–5.
76. B. Huttner, N. Imoto, N. Gisin, T. Mor, *Phys. Rev.*, A 51, 1863, 1995.
77. Huttner, B., J.D. Gautier, A. Muller, H. Zbinden, and N. Gisin, *Phys. Rev.*, A 54, 3783–3789, 1996.
78. W.Y. Hwang, *Phys. Rev. Lett.*, 91, 057901, 2003.
79. H. Weinfurter, *Europhysics Lett.*, 25, 559–564, 1994.
80. I. Marcikic, H. de Riedmatten, W. Tittel, H. Zbinden and N. Gisin, *Nature*, 421, 509–513, 2003; quant-ph/0301178.
81. H. de Riedmatten, I. Marcikic, W. Tittel, H. Zbinden, D. Collins and N. Gisin, *Phys. Rev. Lett.*, 92, 047904, 2004; quant-ph/0309218.
82. Another illustration that research is not deterministic (even less so than quantum physics!) is how the idea of the nowadays so called plug & play configuration came up. During a conference in Corsica on micro-cavities I (NG) attended, there was, as usual, a day off for tourism. The night before the idea came, preventing me from sleeping. I skipped the tourism day and, in my student room in Cargese, wrote the first draft of [4].
83. Note that the mode containing the photon is not a frequency mode since the pulse has a finite spread.
84. And assuming no noise in the detectors.
85. In particular it is impossible to clone perfectly a quantum state.
86. There has been some controversy on the security of faint laser QKD and some people claimed that it is not "unconditional security." However, there is no such thing as "unconditional security." Obvious conditions for secret communication are, for instance, the clear definition of the boundaries of Alice's and Bob's office and, in the case of QKD, the validity of quantum mechanics. For well-defined conditions, Eve's information due to multiphoton pulses can be calculated and reduced to arbitrary low values.

87. Despite their name, polarization maintaining fibers do not maintain any arbitrary polarization state. Only light polarized along one of the fiber's birefringence axes is maintained.

88. The QKD setup uses standard 50% couplers instead of the variable couplers and the switches.

89. A Faraday mirror is a 90-degree Faraday rotator followed by a mirror. The outcoming light is thus orthogonally polarized with respect to the ingoing light. This is true whatever the incoming polarization state. Moreover, if a fiber link is connected to a Faraday mirror, the outcoming light is still orthogonally polarized independently of the polarization evolution inside the fiber [1, 7].

90. Here we assume that the collection efficiency for the photon traveling towards Bob is 1. In practice, a more realistic value is ≈ 0.70.

91. Note that these results are not in contradiction with relativity.

92. To be precise, the assumption of independence only holds when the two pairs are not created simultaneously within their coherence lengths [32], a condition that is easily fulfilled experimentally.

chapter 3

Advanced Quantum Communications Experiments with Entangled Photons

M. Aspelmeyer and A. Zeilinger
University of Vienna and
Austrian Academy of Sciences

H. R. Böhm, A. Fedrizzi, S. Gasparoni*,
M. Lindenthal, G. Molina-Terriza**, A. Poppe,
K. Resch***, R. Ursin and P. Walther
University of Vienna

T.D. Jennewein
Austrian Academy of Sciences

Contents

3.1 Introduction...46
3.2 Advanced Quantum Communication Schemes.....................47
    3.2.1 Scalable Teleportation and Entanglement
        Swapping...47
        3.2.1.1 Entanglement Swapping 48
        3.2.1.2 Scalable Teleportation 48
        3.2.1.3 Long-Distance Quantum Teleportation 51
    3.2.2 Purifying Quantum Entanglement 54
    3.2.3 A Photonic Controlled NOT Gate 55

*Present address: Center for Biomedical Engineering and Physics, Medical University of Vienna, Austria
**Present address: ICFO — Institut de Ciències Fotòniques, Barcelona, Spain
***Present address: Department of Physics, University of Queensland, Brisbane, Australia

3.2.4   Higher Dimensional Entanglement
        for Quantum Communications 60
3.2.5   Entanglement-Based Quantum Cryptography 62
        3.2.5.1   Adopted BB84 Scheme 64
        3.2.5.2   An Entanglement-Based Quantum
                  Cryptography Prototype System 64
3.2.6   Toward a Global Quantum Communication
        Network 68
        3.2.6.1   Free-Space Distribution of Quantum
                  Entanglement 68
        3.2.6.2   Quantum Communications in Space 73
3.3   Conclusion and Outlook 75
Acknowledgments 75
References 76

3.1   Introduction

Quantum communication and quantum computation are novel methods of information transfer and information processing, all fundamentally based on the principles of quantum physics. The performances outdo their classical counterparts in many aspects [1,2]. In almost all quantum communication and quantum computation schemes, quantum entanglement [3] plays a decisive role. In essence, an entangled system can carry all information (e.g., on their polarization properties) only in their correlations, while no individual subsystem carries any information. This leads to correlations that are much stronger than classically allowed [89, 100], which is a powerful resource for information processing. It is therefore important to be able to generate, manipulate, and distribute entanglement as accurately and as efficiently as possible.

Successful demonstrations of quantum communication protocols started with photon experiments in 1992 and include quantum cryptography [4,5], the simultaneous distribution of a cryptographic key that is ultimately secured by the laws of quantum physics; later followed quantum dense coding [6,7], a protocol to double the classically allowed capacity of a communication channel by encoding two bits of information per bit sent, and finally quantum teleportation [8,9], the remote transfer of an arbitrary quantum state between distant locations.

Since these early achievements, the field of quantum communication, or more generally quantum information processing, has very much advanced. New schemes and techniques allow the generation and manipulation of entangled photon pairs and even of four-photon states with much higher efficiency and precision [10,11]. Also, the distances over which entanglement can be distributed are regularly pushed further. Owing to new protocols one can now achieve the successive use of teleported states and also the teleportation of entanglement via entanglement swapping (see Section 3.2.1). An important method for distributing pure entangled states even over noisy channels is entanglement purification (see Section 3.2.2), which is one ingredient for a quantum repeater and is also based on the application of elementary quantum gates such as a controlled NOT (CNOT) gate (see Section 3.2.3). Another promising line of development involves entanglement in higher dimensions, which might allow further advances such as quantum communication with a higher resistance against noise (see Section 3.2.4). A very recent development is the real-world application of entanglement-based quantum cryptography (see Section 3.2.5). This is linked to the research on distributing entanglement over long distances, which aims at the establishment of a quantum communication network (see Section 3.2.6), eventually on a global scale by using satellites.

3.2 Advanced Quantum Communication Schemes

3.2.1 Scalable Teleportation and Entanglement Swapping

Teleportation of quantum states [12] is an intriguing concept within quantum physics and a striking application of quantum entanglement. Besides its importance for quantum computation [13,14], teleportation is at the heart of the quantum repeater [15], a concept eventually allowing the distribution of quantum entanglement over arbitrary distances and thus enabling quantum communication over large distances and even networking on a global scale.

The purpose of quantum teleportation is to transfer an arbitrary quantum state to a distant location, e.g., from Alice to Bob, without transmitting the actual physical object carrying the state. Classically this is an impossible task, since Alice cannot obtain the full information of the state to be teleported without previous knowledge about its preparation. Quantum physics, however, provides a working strategy. Suppose, Alice and Bob share an ancilla entangled pair in advance. Alice then performs a Bell state measurement between the teleportee particle and her shared ancilla, i.e., she projects the two particles into the basis of Bell states. The four possible outcomes of this measurement provide her with two bits of classical information, which is sufficient to reconstruct the initial quantum state at Bob's side. After communicating the classical result to Bob, he can perform one out of four unitary operations to obtain the original state to be teleported. In detail: suppose photon 1, which Alice wants to teleport to Bob, is in a general polarization state $|\chi\rangle_1 = \alpha|H\rangle_1 + \beta|V\rangle_1$ (unknown to Alice), and the pair of photons 2 and 3 shared by Alice and Bob is in the polarization-entangled state $|\Psi^-\rangle_{23}$. This state is one of the four maximally entangled Bell states $|\Psi^\pm\rangle_{ij} = \frac{1}{\sqrt{2}}(|HV\rangle_{ij} \pm |VH\rangle_{ij})$ and $|\Phi^\pm\rangle_{ij} = \frac{1}{\sqrt{2}}(|HH\rangle_{ij} \pm |VV\rangle_{ij})$, where H and V denote horizontal and vertical linear polarizations, and i and j index the spatial modes of the photons. The overall state of photons 1, 2, and 3 can be rewritten as

$$|\Psi\rangle_{123} = |\Psi\rangle_1 |\Psi^-\rangle_{23} = \frac{1}{2}[-|\Psi^-\rangle_{12} (\alpha |H\rangle_3 + \beta |V\rangle_3)$$

$$- |\Psi^+\rangle_{12} (\alpha |H\rangle_3 - \beta |V\rangle_3) \qquad (3.1)$$

$$+ |\Phi^-\rangle_{12} (\alpha |V\rangle_3 + \beta |H\rangle_3)$$

$$+ |\Phi^+\rangle_{12} (\alpha |V\rangle_3 - \beta |H\rangle_3)].$$

It can thus be seen that a joint Bell measurement on photons 1 and 2 at Alice's side, i.e., a projection of particles 1 and 2 onto one of the four Bell states, projects the state of photon 3 at Bob's side into one of the four corresponding states, as shown in Equation (3.1). The outcome of the Bell measurement is totally random (otherwise Alice and Bob could communicate faster than light). However, when knowing Alice's measurement results, Bob can perform a unitary transformation, independent of $|\chi\rangle_1$, on photon 3 and convert its state into the initial state of photon 1.

3.2.1.1 Entanglement Swapping

An important feature of teleportation (also of relevance for long-distance quantum communication) is that it provides no information whatsoever about the state being teleported. This means that an arbitrary unknown quantum state can be teleported. In fact, the quantum state of a teleportee particle does not have to be well defined, and it could thus even be entangled with another photon. A Bell state measurement of two of the photons — one each from two pairs of entangled photons — results in the remaining two photons becoming entangled, even though they have never interacted in the past (see Figure 3.1(a)). This was demonstrated recently by violating a Bell inequality between particles that never interacted with each other [16] (see Figure 3.1(b)). A chain of several entanglement swapping systems [17] can in principle be used to transfer quantum entanglement between distant sites.

3.2.1.2 Scalable Teleportation

A recent result also of relevance for long-distance quantum communication is the first realization of freely propagating teleported qubits [18], which will eventually allow the subsequent use of teleported states. In previous experimental realizations of teleportation with photons, the teleported qubit had to be detected (and thus destroyed) to verify the success of the procedure. This can be avoided by providing, on average, more entangled ancilla pairs than states to be teleported. In the modified teleportation scheme (Figure 3.2), a successful Bell state analysis results in freely propagating individual qubits, which can be used for further cascaded teleportation. In many of our experiments, two independent polarization entangled photon pairs, produced by spontaneous parametric down-conversion (SPDC) with a probability p, are used both for the preparation of the entangled pair $|\Psi^-\rangle_{23}$ (photons 2 and 3) and for the preparation of the initial state to be teleported (photons 1 and 4).

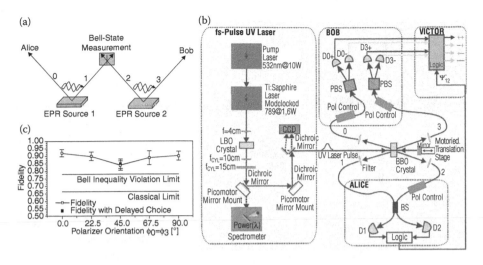

Figure 3.1 (a) Scheme for entanglement swapping, i.e., the teleportation of entanglement. Two entangled pairs 0, 1 and 2, 3 are produced by two entangled photon sources (EPR). One particle from each of the pairs is sent to two separated observers; say 0 is sent to Alice and 3 to Bob. 1 and 2 become entangled through a Bell state measurement, by which 0 and 3 also become entangled. This requires the entangled qubits 0 and 3 neither to come from a common source nor to have interacted in the past. (b) Experimental setup for the demonstration of teleportation and entanglement swapping using pairs of polarization entangled photons. The two entangled photon pairs were produced by down-conversion in barium borate (BBO), pumped by femtosecond UV laser pulses traveling through the crystal in opposite directions. After spectral filtering, all photons were collected in single-mode optical fibers for further analysis and detection. Single-mode fibers offer the benefit that the photons remain in a perfectly defined spatial mode allowing high-fidelity interference. In order to optimize the temporal overlap between photon 1 and 2 in the beam splitter, the UV mirror was mounted on a motorized translation stage. Photons 0 and 3 were sent to Bob's two-channel polarizing beamsplitters for analysis, and the required orientation of the analyzers was set with polarization controllers in each arm. All photons were detected with silicon avalanche photodiodes, with a detection efficiency of about 40%. Alice's logic circuit detected coincidences between detectors D1 and D2. (c) Experimental violation of Bell's inequalities from particles that never interacted with each other obtained through correlation measurements between photons 0 and 3, which is a lower bound for the fidelity of the teleportation procedure (from [16]). ϕ_0 (ϕ_3) is the setting of the polarization analyzer for photon 0 (photon 3) and $\phi_0 = \phi_3$. The minimum fidelity of 0.84 is well above the classical limit of 2/3 and also above the limit of 0.79 necessary for violating Bell's inequality.

Photon 4 acts as a trigger to indicate the presence of photon 1. If one pair of photons is emitted in each of the pairs of modes 1-4 and 2-3, a threefold coincidence of T-D1-D2 is sufficient to guarantee a successful teleportation.

However, owing to the probabilistic nature of SPDC, two photon pairs are both emitted into modes 1-4 with the same probability p^2 as for a

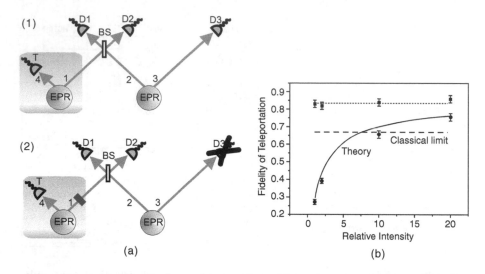

Figure 3.2   (a) (1) Original qubit teleportation scheme using polarization-entangled photon pairs based on spontaneous parametric down-conversion (EPR sources); (2) Freely propagating teleported qubits. (b) Conditional fidelities (squares) and nonconditional fidelities (circles) obtained in 45° teleportation for different attenuation $1/\gamma$. With increasing attenuation of mode 1, an increase of nonconditional fidelities is observed while the conditional ones remain constant. For $\gamma = 0.05$, the classical limit $2/3$ is clearly overcome.

successful teleportation. This will also lead to threefold coincidences of T-D1-D2, but in this case no teleportation occurs, as mode 3 is simply empty. To ensure a successful teleportation, it has been necessary to confirm the presence of photon 3 by actually detecting it (Figure 3.2). For this reason, the original Innsbruck experiment [9], the first experimental demonstration of quantum teleportation, has been called a "postselected" one, while probably the word "conditional" would be more appropriate, as detection of photon 3 does not depend on its state. In the new protocol, an unbalanced two-photon interferometer is used to make a detection at Bob's side obsolete and therefore allow for a free propagation of the teleported qubits. The number of unwanted D1-D2 coincidence counts is reduced by attenuating beam 1 by a factor of γ while leaving the modes 2-3 unchanged. Then a threefold coincidence D1-D2-T will occur with probability p^2 owing to successful teleportation, and with significantly lower probability γp^2 there will be a spurious coincidence. Thus the probabilty for a successful teleportation event (conditioned on D1-D2-T) will scale with $1/(1 + \gamma)$, and for sufficiently low γ it will not be necessary anymore to detect the teleported photon 3; a freely propagating teleported beam of qubits emerges.

To demonstrate experimentally nonconditional teleportation, we inserted a series of neutral density filters in mode 1 and showed that the probability of having a successful teleportation conditioned on a threefold coincidence

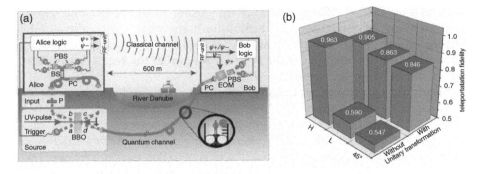

Figure 3.3   (a) Sketch of the two laboratories located on either side of the Danube River (from [23]). The laboratories were located in two sewage water system buildings owned by the city of Vienna. The faster classical channel (microwave) and the slower quantum channel (fiber) are shown above and underneath the Danube. The vertical separation of the two channels is about 40 m. (b) Fidelity of the teleported states with and without active switching.

of D1, D2, and T increases with decreasing γ. The corresponding fidelities for conditional (fourfold detection) and nonconditional (threefold detection D1-D2-T) teleportation are shown in Figure 3.2b. We were able to demonstrate the preparation of a freely propagating teleported quantum state with high (nonconditional) fidelity of $0.85 \perp 0.02$, i.e., well above the classical limit.* The possibility of letting the teleported qubit travel freely in space, together with the high experimental visibility obtained, is a fundamental step in the direction of the realization of long-distance quantum communication. Further protocols, such as entanglement purification, are then needed to overcome decoherence in long-distance quantum channels (see Section 3.2.2).

3.2.1.3   Long-Distance Quantum Teleportation

Teleportation is the basis for the quantum repeater [15], which allows distributing quantum entanglement over long distances. Also, quantum teleportation over longer distances will be needed for realizing quantum network schemes involving several parties [22] and naturally for interconnecting devices utilizing quantum computational algorithms. As a next step toward a full-scale implementation of a quantum repeater, we have realized a large-scale implementation of teleportation [23] of photon qubits in an outdoor environment. The two laboratories involved, Alice and Bob, are separated by 600 m across the Danube River in Vienna (see Figure 3.3). Additionally, while it has been shown that systems based on linear optical elements can only determine two of the four Bell states perfectly [24,7,25], our system achieves the

---

*Other nonconditional quantum teleportation experiments have been performed with continuous variables [19] and, only recently, with ions [20,21]. However, the most suitable systems for long-distance transmission are currently photons.

optimal teleportation efficiency of 50% when using linear optics alone. This is realized by an active feed-forward technique, namely by detecting two of the four Bell states on the transmitter site, Alice, and correspondingly switching the unitary transformation for the receiver photon at the receiver site, Bob, with a fast electro optic modulator (EOM).

An important feature of our experiment was the implementation of an optimized Bell state analyzer (BSA) capable of detecting two of the four Bell states on Alice's side, and the implementation of an "actively switched" unitary transformation on Bob's side triggered by the outcome of Alice's Bell state measurement (BSM). Alice's BSM result was sent to Bob via a microwave link (2.4 GHz) above the Danube River. The classical channel transmitted one bit of information about Alice's BSM outcome (Ψ^- or Ψ^+). This signal traveled the distance of 600 m almost at the vacuum speed of light, which took about 2 μs. Additionally, delays in the detectors and in several signal stages of the transceivers introduced an extra delay of 0.6 μs. However, since the speed of light in a fiber is approximately 2/3 of the vacuum speed of light, the entangled photon d traveled the 800 m fiber from Alice to Bob in about 4 μs. Therefore, the information on Alice's BSA outcome arrived at Bob's laboratory approximately 1.4 μs before the arrival of photon d. This provided Bob with sufficient time to set an EOM to apply the birefringent phase shift of 0 or π between the $|H\rangle$ and $|V\rangle$ optical light modes on the received photon d. When Alice's BSM result was a Ψ^-, then Bob left the EOM at the idle voltage (i.e., the EOM introduces no phase shift and the teleported state remains unchanged), and when it was a Ψ^+, then Bob applied the activation voltage (i.e., the EOM introduced a π phase shift between the horizontal and vertical polarization) just before the photon passed through the EOM. In both cases, Bob eventually obtains an exact replica of the initial teleportee state (see Equation (3.1)).

The EOM was a KDP Pockels cell, which achieved the phase shift of π with an accuracy of 1:200 at a voltage of 3.4 kV. The timing information of the received classical signal was delayed with a digital delay generator, and when a Ψ^+ was received, then the EOM was triggered with a 100 ns pulse with a rise time of 20 ns. Additionally, Bob used a logic circuit to count only those detections of his photons that arrived at the expected times within a coincidence window of ±10 ns. Note that without operation of the EOM, Bob observes only a completely mixed polarization for a $|45°\rangle$ and a circular polarization input state. When teleporting polarizations along H or V, the transformation is irrelevant, since the EOM phase shift does not affect these states. The full teleportation protocol was demonstrated by teleporting distinct linear polarization states and circular polarization states (see Figure 3.4b). The classical fidelity limit of 2/3 [26] is clearly surpassed by our observed fidelities of around 0.85. This is a step toward a full-scale implementation of a quantum repeater to achieve shared pure entanglement between arbitrarily separated quantum communication partners. Such a quantum repeater requires quantum teleportation, purification of entanglement, and quantum memories. Recent results [27,28] indicate the advance of quantum memory, which might be suitable for future quantum communication networks.

(a)

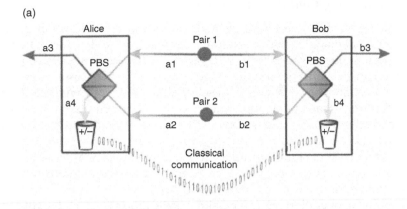

(b)

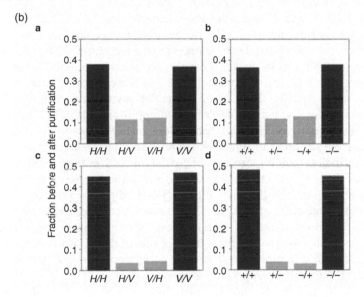

Figure 3.4 (a) Scheme for entanglement purification of polarization-entangled qubits [30]. Two shared pairs of an ensemble of equally mixed entangled states ρ_{AB} are fed into the input ports of polarizing beam splitters that substitute the bilateral CNOT operation necessary for a successful purification step. Alice and Bob keep only those cases where there is exactly one photon in each output mode. This can happen only if no bit-flip error occurs over the channel. Finally, to obtain a larger fraction of the desired pure (Bell) state they perform a polarization measurement in the $|\pm\rangle$ basis in modes a_4 and b_4. Depending on the results, Alice performs a specific operation on the photon in mode a_3. After this procedure, the remaining pair in modes a_3 and b_3 will have a higher degree of entanglement than the two original pairs. (b) Experimental results. **a** and **b** show the experimentally measured fractions both in the H/V and in the $+/-$ bases for the original mixed state. **c** and **d** show the measured fractions of the purified state in the modes a_3 and b_3 both in the H/V and in the $+/-$ bases. Compared with the fractions in **a** and **b**, the experimental results shown in **c** and **d** both together confirm the success of entanglement purification.

3.2.2   Purifying Quantum Entanglement

Owing to unavoidable decoherence in the quantum communication channel, the quality of entangled states generally decreases with the channel length. Entanglement purification schemes [29] allow two spatially separated parties to convert an ensemble of partially entangled states (which result from transmission through noisy channels) to a set of almost perfectly entangled states by performing local unitary operations and measurements on the shared pairs, and coordinating their actions with a classical channel. One thus simulates a noiseless quantum channel by a noisy one, supplemented by local actions and classical communication. In a recent experiment, entanglement purification could be demonstrated for the first time experimentally for mixed polarization-entangled two-particle states [30].

The crucial operation for a successful purification step is a bilateral conditional NOT (CNOT) gate, which effectively detects single bit-flip errors in the channel by performing local CNOT operations (see Section 3.2.3) at Alice's and Bob's side between particles of shared entangled states. The outcome of these measurements can be used to correct for such errors and eventually leads to a less noisy quantum channel [29]. For the case of polarization entanglement, such a parity check on the correlations can be performed in a straightforward way by using polarizing beamsplitters (PBS) [31] that transmit horizontally polarized photons and reflect vertically polarized ones.

Consider the situation in which Alice and Bob have established a noisy quantum channel, i.e., they share a set of equally mixed, entangled states ρ_{AB}. At both sides the two particles of two shared pairs are directed into the input ports a_1, a_2 and b_1, b_2 of a PBS (see Figure 3.2). Only if the entangled input states have the same correlations, i.e., they have the same parity with respect to their polarization correlations, will the four photons exit in four different outputs (four-mode case), and a projection of one of the photons at each side will result in a shared two-photon state with a higher degree of entanglement. All single bit-flip errors are effectively suppressed.

For example, they might start with the mixed state $\rho_{AB} = F \cdot |\Phi^+\rangle\langle\Phi^+|_{AB}$ $+ (1 - F) \cdot |\Psi^-\rangle\langle\Psi^-|_{AB}$, where $|\Phi^+\rangle = (|HH\rangle + |VV\rangle)$ is one of the four maximally entangled Bell states. Then only the combinations $|\Phi^+\rangle_{a_1,a_2} \otimes$ $|\Phi^+\rangle_{b_1,b_2}$ and $|\Psi^-\rangle_{a_1,a_2} \otimes |\Psi^-\rangle_{b_1,b_2}$ will lead to a four-mode case, while $|\Phi^+\rangle_{a_1,a_2}$ $\otimes |\Psi^-\rangle_{b_1,b_2}$ and $|\Psi^-\rangle_{a_1,a_2} \otimes |\Phi^+\rangle_{b_1,b_2}$ will be rejected. Finally, a projection of the output modes a_4, b_4 into the basis $|\pm\rangle = \frac{1}{\sqrt{2}}(|H\rangle \pm |V\rangle)$ will create the pure states $|\Phi^+\rangle_{a_3,b_3}$ with probability $F' = F^2/[F^2 + (1 - F)^2]$ and $|\Psi^+\rangle_{a_3,b_3}$ with probability $1 - F'$, respectively. The fraction F' of the desired state $|\Phi^+\rangle$ becomes larger for each purification step if $F > \frac{1}{2}$. In other words, the new state ρ'_{AB} shared by Alice and Bob after the bilateral parity operation demonstrates an increased fidelity with respect to a pure, maximally entangled state. This is the purification of entanglement.

Typically, in the experiment, one photon pair of fidelity 92% could be obtained from two pairs, each of fidelity 75%. Also, although only bit-flip errors in the channel have been discussed, the scheme works for any general mixed state, since any phase-flip error can be transformed to a bit-flip by a rotation in a complementary basis. In these experiments, decoherence is overcome to the extent that the technique would achieve tolerable error rates for quantum repeaters in long-distance quantum communication based only on linear optics and polarization entanglement.

Purification not only provides a way to implement long-distance quantum communication but also plays an important role in fault-tolerant quantum computation. Quantum error correction [32,33] allows a universal quantum computer to be operated in a fault tolerant way [34,35]. However, in order for quantum repeaters and quantum error correction schemes to work, there are stringent requirements on the precision of logic operations between two qubits. While the tolerable error rate of logic gates in quantum repeaters is of the order of several percent [15], that in quantum error correction is of the order of 10^{-4} to 10^{-5}, still far beyond experimental feasibility. Fortunately, a recent study shows that entanglement purification can also be used to increase the quality of logic operations between two qubits by several orders of magnitude [36]. In essence, this implies that the threshold for tolerable error in quantum computation is within reach using entanglement purification and linear optics. Our experiments achieved an accuracy of local operations at the PBS of about 98%, or equivalently an error probability of 2%. Together with the high fidelity achieved in the latest photon teleportation experiments, the present purification experiment implies that the threshold of tolerable error rates in quantum repeaters can be well fulfilled. This opens the door to realistic long-distance quantum communication. On the other hand, with the help of entanglement purification the strict accuracy requirements of the gate operations for fault-tolerant quantum computation are also reachable, for example, within the frame of linear optics quantum computation [37].

3.2.3   A Photonic Controlled NOT Gate

As can be seen above, advanced quantum communication protocols such as entanglement purification may require nontrivial manipulation of qubits, e.g., bilateral parity checks on pairs of photons (see Section 3.2.2). The underlying operations are typically elementary quantum gates, which are also used for universal quantum computation. Well-known examples of such gates are the controlled NOT (CNOT) and controlled phase (CPhase) operations. The crucial trait of these gates is that they can change the entanglement between qubits. A CNOT gate flips the value of the target bit if and only if the control bit has the logical value 1 (see below). Entanglement can now be created between two independent input qubits if the control bit is in a superposition of 0 and 1. On the other hand, any Bell state that is fed into a CNOT gate will result in distinct separable output states that make it possible to distinguish all four Bell states deterministically.

In previous experiments [38,39,40] destructive linear optical gate operations have been realized. However, such schemes necessarily destroy the output state and are hence not classically feed-forwardable, i.e., they do not

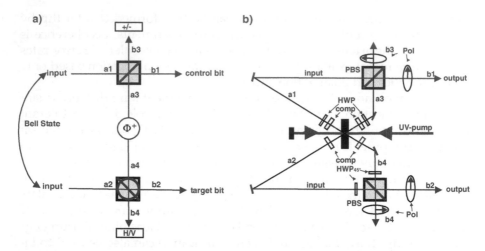

Figure 3.5 (a) The scheme to obtain a photonic realization of a CNOT gate with two independent qubits. The qubits are encoded in the polarization of the photons. The scheme makes use of linear optical components, polarization entanglement, and postselection. When one and only one photon is detected at the polarization sensitive detectors in the spatial modes b_3 and b_4, the scheme works as a CNOT gate. (b) The experimental setup. A type II spontaneous parametric down-conversion is used both to produce the ancilla pair (in the spatial modes a_3 and a_4) and to produce the two input qubits (in the spatial modes a_1 and a_2). In this case, initial entanglement polarization is not desired, and it is destroyed by making the photons go through polarization filters, which prepare the required input state. Half-wave plates (HWP) have been placed in the photon paths in order to rotate the polarization; compensators (comp) are able to nullify the birefringence effects of the nonlinear crystal. Overlap of the wavepackets at the PBSs is assured through spatial and spectral filtering.

allow scalable quantum computation. This section will discuss the realization of a CNOT gate, which operates on two polarization qubits carried by independent photons and which satisfies the feed-forwardability criterion [41]. The scheme, shown in Figure 3.5, was first proposed by Franson et al. [42]. It performs a CNOT operation on the input photons in spatial modes a_1 and a_2; the output qubits are contained in spatial modes b_1 and b_2. The ancilla photons in the spatial modes a_3 and a_4 are in the maximally entangled Bell state $|\phi^+\rangle_{a_3,a_4} = \frac{1}{\sqrt{2}}(|H\rangle_{a_3}|H\rangle_{a_4} + |V\rangle_{a_3}|V\rangle_{a_4})$.

In the following, $|H\rangle$ (a horizontally polarized photon) and $|V\rangle$ (a vertically polarized photon) will denote our logical "0" and "1". The CNOT operation for qubits encoded in polarization can be written as $|H\rangle_c|H\rangle_t \rightarrow |H\rangle_c|H\rangle_t$, $|H\rangle_c|V\rangle_t \rightarrow |H\rangle_c|V\rangle_t$, $|V\rangle_c|H\rangle_t \rightarrow |V\rangle_c|V\rangle_t$, $|V\rangle_c|V\rangle_t \rightarrow |V\rangle_c|H\rangle_t$, where the indices c and t denote the control and target qubit.

The scheme works in those cases where one and only one photon is found in each of the modes b_3 and b_4. It combines two simpler gates, namely the destructive CNOT and the quantum encoder. The first gate can be seen in the lower part of Figure 3.5 and comprises a polarizing beam splitter (PBS2) rotated by 45° (the rotation is represented by the circle drawn inside the symbol of the PBS), which works as a destructive CNOT gate on the polarization qubits, as was experimentally demonstrated in [42]. The upper part, comprising the entangled state and the PBS1, is meant to encode the control bit in the two channels a_4 and b_1.

Owing to the PBS operation, which transmits horizontally polarized photons and reflects vertically polarized ones, the successful detection of the state $|+\rangle$ at the port b_3 postselects the following transformation of the arbitrary input state in a_1: $\alpha|H\rangle_{a_1} + \beta|V\rangle_{a_1} \rightarrow \alpha|H\rangle_{a_4}|H\rangle_{b_1} + \beta|V\rangle_{a_4}|V\rangle_{b_1}$. The control bit is thus encoded in a_4 and in b_1. The photon in a_4 serves as the control input to the destructive CNOT gate and will be destroyed, while the second photon in b_1 serves as the output control qubit.

For the gate to work properly, one has to demonstrate that the most general input state,

$$|\Psi\rangle^{in}_{a_1,a_2} = |H\rangle_{a_1}(\alpha_1|H\rangle_{a_2} + \alpha_2|V\rangle_{a_2}) + |V\rangle_{a_1}(\alpha_3|H\rangle_{a_2} + \alpha_4|V\rangle_{a_2}),$$

can be converted to the output state,

$$|\Psi\rangle^{out}_{b_1,b_2} = |H\rangle_{b_1}(\alpha_1|H\rangle_{b_2} + \alpha_2|V\rangle_{b_2}) + |V\rangle_{b_1}(\alpha_3|V\rangle_{b_2} + \alpha_4|H\rangle_{b_2}).$$

Let us consider first the case where the control photon is in the logical zero or horizontally polarized. The control photon will then travel undisturbed through the PBS, arriving in the spatial mode b_1. As required, the output photon is $|H\rangle$. In order for the scheme to work, a photon needs to arrive also in mode b_3: given that the input photon is already in mode b_1, the additional photon will necessarily be provided by the EPR pair and is $|H\rangle$ after transmission through PBS1. We know that the photons in a_3 and a_4 are entangled, so the photon in a_4 is also in the horizontal polarization state. For a $|H\rangle$ ($|V\rangle$) target photon, taking into account the 45° rotation of the polarization on the paths a_2, a_4 due to the half-wave plates, the input at PBS2 will then be in the state $|+\rangle_{a_2}|+\rangle_{a_4}(|+\rangle_{a_2}|-\rangle_{a_4})$. This state will give rise, with a probability of 50%, to the state where two photons go through the PBS2, namely $|\phi^\pm\rangle_{b_2,b_4} = \frac{1}{\sqrt{2}}(|H\rangle_{b_2}|H\rangle_{b_4} \pm |V\rangle_{b_2}|V\rangle_{b_4})$. After the additional rotation of the polarization and after the subsequent change to the H/V basis (where the measurement will be performed), this state acquires the form $|\phi^+\rangle_{b_2,b_4} = \frac{1}{\sqrt{2}}(|H\rangle_{b_2}|H\rangle_{b_4} + |V\rangle_{b_2}|V\rangle_{b_4})$ $(|\psi^+\rangle_{b_2,b_4} = \frac{1}{\sqrt{2}}(|H\rangle_{b_2}|V\rangle_{b_4} + |V\rangle_{b_2}|H\rangle_{b_4}))$.

The expected result, $|H\rangle$($|V\rangle$), in the mode b_2 is found for the case where the photon in b_4 is horizontally polarized. We can see in a similar way that the gate works also for the cases where the control photon is vertically polarized or is polarized at 45°.

The experimental setup for the CNOT gate is shown in Figure 3.5. An ultraviolet pulsed laser, centered at a wavelength of 398 nm, with pulse duration 200 fs and a repetition rate of 76 MHz, impinges on a nonlinear BBO crystal [43], in which it produces probabilistically the first photon pair in the spatial modes a_1 and a_2. They serve as input qubits to the gate. The UV laser is reflected back by the mirror M1 and, on passing through the crystal a second time, produces the entangled ancilla pair in spatial modes a_3 and a_4.

Half-wave plates and nonlinear crystals in the paths provide the necessary birefringence compensation, while the same half-wave plates are used to adjust the phase between the down-converted photons (i.e., to produce the state ϕ^+) and to implement the CNOT gate.

We then superpose the two photons at Alice's (Bob's) side in the modes a_1, a_3 (a_2, a_4) at a polarizing beam splitter PBS1 (PBS2). The indistinguishability between the overlapping photons is improved by introducing narrow bandwidth (3 nm) spectral filters at the outputs of the PBSs and monitoring the outgoing photons by fiber-coupled detectors. The single-mode fiber couplers guarantee good spatial overlap of the detected photons; the narrow bandwidth filters stretch the coherence time to about 700 fs, substantially larger than the pump pulse duration [44]. The temporal and spatial filtering process effectively erases any possibility of distinguishing the photon pairs and therefore allows two-photon quantum interference.

The described CNOT scheme is nondestructive, i.e., the output photons can travel freely in space and may be further used in quantum communication protocols. This is achieved by detecting one and only one photon in modes b_3 and b_4. Since photon-number resolving detectors are not yet readily available at this wavelength, we implement a fourfold coincidence detection to confirm that photons actually arrive in the output modes b_1 and b_2.

To demonstrate experimentally the working operation of the CNOT gate, we first verify the CNOT truth table for input qubits in the computational basis states $|H\rangle|H\rangle$, $|H\rangle|V\rangle$, $|V\rangle|H\rangle$, and $|V\rangle|V\rangle$. Figure 3.6(b) compares the count rates for all 16 possible combinations. We then show that the gate also works for a superposition of input states. The special case in which the control input is a 45° polarized photon and the target qubit is an $|H\rangle$ photon is particularly interesting: we expect that the state $|+\rangle_{a_1}|H\rangle_{a_2}$ evolves into the maximally entangled state $|\phi^+ b_1, b_2 = \frac{1}{\sqrt{2}}(|H\rangle_{b_1}|H\rangle_{b_2} + |V\rangle_{b_1}|V\rangle_{b_2})$. We prepare $|+\rangle_{a_1}|H\rangle_{a_2}$ as the input state; first we measure the count rates of the four combinations of the output polarization ($|H\rangle|H\rangle, \ldots, |V\rangle|V\rangle$) and then after going to the $|\pm\rangle$ linear polarization basis an Ou–Hong–Mandel interference measurement is possible; this is shown in Figure 3.6.

On the other hand, the same CNOT operation can be used to identify Bell states when they are used as input states [45]. For this procedure, the gate performs an operation that transforms each of the entangled Bell states into well-defined but different separable states, which are simple to distinguish. When a Bell state enters a CNOT gate in modes a_1 and a_2, the gate operation can be described by

$$|\phi^{\pm}\rangle_{a_1,a_2} \rightarrow |\pm\rangle_{b_1}|H\rangle_{b_2}$$

$$|\psi^{\pm}\rangle_{a_1,a_2} \rightarrow |\pm\rangle_{b_1}|V\rangle_{b_2}.$$

Figure 3.6 shows the count rates of all 16 possible combinations (four different inputs and four different outputs). They clearly confirm the successful implementation of the Bell state analyzer. The fidelity of each Bell state analysis is $F_{\phi^+} = (0.75 \pm 0.05)$, $F_{\phi^-} = (0.79 \pm 0.05)$, $F_{\psi^+} = (0.79 \pm 0.05)$,

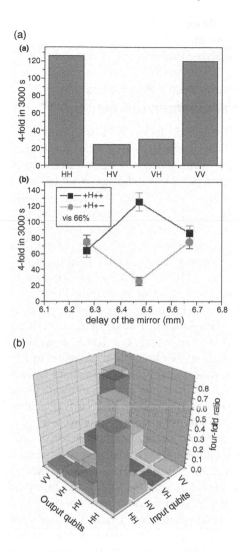

Figure 3.6 (a) Demonstration of the ability of the CNOT gate to transform a separable state into an entangled state. In (**a**), the coincidence ratio between the different terms $|H\rangle|H\rangle, \ldots, |V\rangle|V\rangle$ is measured, proving that the birefringence of the PBS has been sufficiently compensated; in (**b**) the superposition between $|H\rangle|H\rangle$ and $|V\rangle|V\rangle$ is proved to be coherent, by showing via the Ou–Hong–Mandel dip at $45°$ that the desired $|+\rangle$ state of the target bit emerges much more often than the spurious state $|-\rangle$. The fidelity is of $(81 \pm 2)\%$ in the first case and $(77 \pm 3)\%$ for the second. (b) Experimental demonstration of the optical Bell state analyzer. Fourfold coincidences for all possible 16 combinations of the inputs and outputs are shown. Each of the four different polarization-encoded Bell states is transformed into a distinguishable separable state $|\phi^{\pm}\rangle_{a_1,a_2} \to |\pm\rangle_{b_1}|H\rangle_{b_2}$ and $|\psi^{\pm}\rangle_{a_1,a_2} \to |\pm\rangle_{b_1}|V\rangle_{b_2}$. Each input state was measured for 1800 seconds at each of the four different polarizer settings. The fidelity is, on average, 77%.

$F_{\psi^-} = (0.75 \pm 0.05)$, without subtracting background of any kind. The incorrect outcomes originate mainly from incomplete suppression of the double pair emission and imperfections in the PBS operation.

3.2.4   Higher Dimensional Entanglement for Quantum Communications

In classical communication protocols it is not unusual to send the information encoded not only in 0's and 1's but also in a higher number of levels. For example, when the information is encoded in phase-modulated electrical signals, coding two bits per phase change doubles the number of bits per second. This is called two-level coding. This method is suitable, for example, for 2400 bps modems (CCITT V.26). Encoding more bits per phase change increases the number of bits per second but, assuming a constant noise, decreases the signal-to-noise ratio (SNR). The right choice of level coding per physical information carrier is a complicated engineering problem which, besides the speed and the SNR, includes the elaboration of new and efficient communication protocols for higher level encoding.

As we have already reviewed in previous chapters, quantum communication and quantum computation protocols usually encode the information in two-dimensional quantum systems, better known as qubits. Nevertheless, there are ways of enlarging the available dimension of the quantum information carrier. A system that is completely described by n different orthogonal vectors is called a qunit. In the same way as in their classical counterpart, the use of qunits increases the information rate, but surprisingly enough the system is also more resistant to noise. For example, entangled qunits can violate Bell's inequalities more than their two-dimensional counterpart, protecting in this way the nonlocal quantum correlations against noise. Also, a quantum cryptography protocol using qunits is usually more secure against noise than those protocols based on qubits [46,47,48]. On the other hand, there are a series of protocols in quantum communication that are designed specifically for being implemented in higher dimensional spaces [49,50,51]. On a more fundamental level, higher dimensional Hilbert spaces provide novel counterintuitive examples of the relationship between quantum information and classical information, which cannot be found in two-dimensional systems [52,53,54].

Encoding qunits with photons has been experimentally demonstrated using interferometric techniques such as time-bin schemes [55] and superpositions of spatial modes [56]. Up to now, the only noninterferometric technique of encoding qunits in photons is using their orbital angular momentum or, equivalently, their transversal modes [57,58]. Orbital angular momentum modes usually contain dark spots that regularly exhibit phase singularities.

The orbital angular momentum of light has already been used to entangle and to concentrate entanglement of two photons [57,59]. This entanglement has also been shown to violate a two particle three-dimensional Bell inequality [60]. There have been proposals of some experimental techniques to engineer entangled qunits in photons [58,61,62]. Here we will discuss a general scheme for a quantum communication protocol based on the orbital angular momentum of light [63]. This scheme has already been succesfully used for the experimental realization of a quantum coin tossing protocol [64].

In a general communication scheme, prior to the sharing of information, the two parties, say Alice and Bob, have to define a procedure that will assure that the signal sent by one party is properly received by the other. Usually, this scheme works as follows. First, Alice prepares a signal state she wants to send. Bob will measure it and communicate the result to Alice, who will correct the parameters of her sending device following Bob's indications. This process will be repeated until the two parties adjust the corresponding devices. After this step is completed, Alice can safely assume that any subsequent signal which is sent is properly received by Bob.

Using pairs of photons entangled in orbital angular momentum, we can prepare any qutrit state, transmit it, and measure it. The preparation is done by projecting one of the two photons onto some desired state. This projects the second photon nonlocally onto a corresponding state. This state may be transmitted to Bob and finally measured by him. The measurement employs tomographic reconstruction. This last step is usually a technically demanding problem, inasmuch as it needs the implementation and control of arbitrary transformations in the quantum system's Hilbert space.

The experimental setup we used is shown in Figure 3.7. A 351 nm wavelength argon-ion laser pumps a 1.5-mm-thick BBO (β-barium-borate) crystal cut for Type I phase matching conditions. The crystal is positioned so as to produce down-converted pairs of equally polarized photons at a wavelength of 702 nm emitted at an angle of 4° off the pump direction. These photons are directly entangled in the orbital angular momentum degree of freedom. Alice can manipulate one of the down-converted photons while the other is sent to Bob. Before being detected, Bob's photon traverses two sets of holograms.

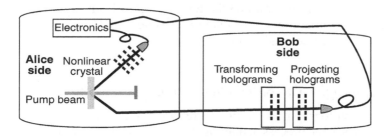

Figure 3.7 Experimental setup from [63]. A 351 nm wavelength laser pumps a BBO crystal. The two generated 702 nm down-converted photons are sent to Alice's and Bob's detectors, respectively. Before being detected, each photon propagates through a set of holograms. Each photon was coupled into single-mode fibers and directed to detectors based on avalanche photodiodes operating in the photon counting regime.

Each set consists of one hologram with charge $m = 1$ and another with charge $m = -1$. The first set of holograms provides the means of a transformation in the three-dimensional space expanded by the states $|-1\rangle$, $|0\rangle$, and $|1\rangle$. The second set, together with a single-mode fiber and a detector, acts as a projector onto the three different basis states. All these elements are Bob's receiving device. Alice's photon also traverses a set of holograms, which, together with the source and the detector on Alice's side, act as Alice's sending device. Whenever Alice detects one photon, the transmission of a photon to Bob is initiated. Due to the quantum correlations between the entangled photons, Alice can radically control the state of the photon sent to Bob. In order to adjust their respective devices properly, Bob has to perform a tomographic measurement of the received state and classically to communicate his result to Alice.

In Figure 3.8 we present three examples of qutrits that were received by Bob and remotely prepared by Alice. All of them were found to be very nearly pure states, their largest eigenvalues and corresponding eigenvectors being (a) $\lambda_{max} = 0.99$, $|e_{max}\rangle = 0.68|0\rangle + 0.71|1\rangle - 0.14|-1\rangle$; (b) $\lambda_{max} = 0.99$, $|e_{max}\rangle = 0.65|0\rangle + 0.53\exp(-i0.26\pi)|1\rangle + 0.55\exp(-i0.6\pi)|-1\rangle$; (c) $\lambda_{max} = 0.99$, $|e_{max}\rangle = 0.58|0\rangle + 0.58\exp(-i0.05\pi)|1\rangle + 0.58\exp(-i0.89\pi)|-1\rangle$. From these examples it is shown that besides the relative intensities, Alice could also control the relative phases of the states sent. Other reconstructed qutrits (not presented in Figure 3.8) showed an effective suppression of the $|0\rangle$ mode through destructive interference from the two holograms. The result was $\lambda_{max} = 0.97$, $|e_{max}\rangle = 0.26|0\rangle + 0.68\exp(i0.11\pi)|1\rangle + 0.68\exp(-i0.21\pi)|-1\rangle$. As can be deduced from the maximum eigenvalue of all the data, the purity of the reconstructed states was larger than 97%. By direct comparison of the measured data and the data estimated by the reconstructed matrix, the error was comparable to the statistical Poissonian noise, which demonstrates the reliability of the tomography.

The method presented establishes a point-to-point communication protocol in a three-dimensional alphabet. Using the orbital angular momentum of photons, we can implement the three basic tasks inherent in any communication or computing protocol: preparation, transmission, and reconstruction of a qutrit. This communication scheme has already been experimentally implemented in a quantum coin tossing experiment [64], which is an original cryptographic protocol using qutrits.

3.2.5 Entanglement-Based Quantum Cryptography

Quantum cryptography is the first technology in the area of quantum information that is in the process of making the transition from purely fundamental scientific research to an industrial application. In the last three years, several companies have started developing quantum cryptography prototypes, and the first products have hit the market. Up to now, these commercial products were all based on various faint-pulse implementations of the BB84 protocol [5,65,66].

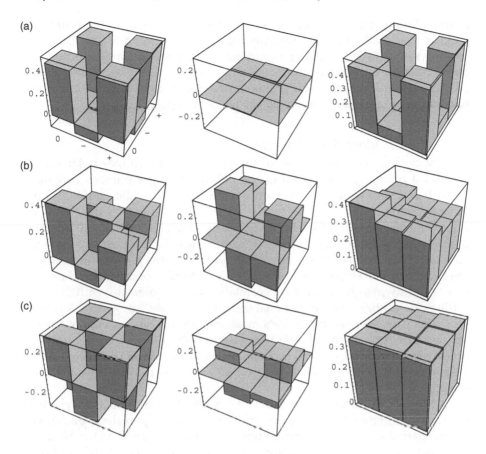

Figure 3.8 Results of quantum state tomography applied to three different re-
motely prepared states of Bob's qutrits: (a) $0.68|0\rangle + 0.71|1\rangle - 0.14|-1\rangle$; (b) $0.65|0\rangle +$
$0.53\exp(-i0.26\pi)|1\rangle + 0.55\exp(-i0.6\pi)|-1\rangle$;   (c) $0.58|0\rangle + 0.58\exp(-i0.05\pi)|1\rangle +$
$0.58\exp(-i0.89\pi)|-1\rangle$. Left and middle panels show real and imaginary parts of
the reconstructed density matrices; right panels show the absolute values of those
elements for better comparison of how large are the contributions of the three basic
states. From the results it is shown that Alice can control both the relative amplitudes
and the phases of the sent states.

The use of entanglement provides a superior approach to quantum cryptography and was first proposed by Ekert [67]. One of the main conceptional
advantages over single-photon quantum cryptography is the inherent randomness in the results of a quantum-mechanical measurement on an entangled system leading to purely random keys. Furthermore, the use of entangled pairs eliminates the need for a deterministic single photon source,
because a pure entangled photon state consists, by definition, of exactly two
photons that are sent to different recipients. Multiple-pair emissions are inherently rejected by the protocol, in contrast to the faint-pulse case, where a

beam-splitting attack might be successful.* Additionally, high-intensity sources would allow longer transmission paths compared to single-photon based systems [69,70]. Another important advantage over single-photon systems is that the photon pair source is immune to tampering by an illegitimate party. Any manipulation at the photon source can be detected by the communicating parties and communication can be stopped.

3.2.5.1   Adopted BB84 Scheme

A very elegant implementation of the BB84 scheme utilizes polarization-entangled photon pairs instead of the polarized single photons originally used. This is very similar to the Ekert scheme [67] when Alice and Bob chose different settings of their analyzers for their measurements of the entangled photons. As opposed to the Ekert scheme, in which both Alice and Bob randomly vary their analyzers between three settings, the adopted BB84 scheme uses only two analyzer states, namely $0°$ and $45°$. If they share, for example, the entangled Bell singlet state $|\Psi^-\rangle$, Alice's and Bob's polarization measurements will always give perfect anticorrelations if they measure with the same settings, no matter whether the analyzers are both at $0°$ or at $45°$. A way to view this is to assume that Alice's measurement on her particle of the entangled pair projects the photon traveling to Bob onto the orthogonal state of the one observed by Alice. So the photons transmitted to Bob are polarized in one of the four polarizations $0°$, $45°$, $90°$ and $135°$, as with the BB84 scheme.

After a measurement run, when Alice and Bob independently collect photons for a certain time, they communicate over an open classical channel. By comparing a list of all detection times of photons registered by Alice and Bob, they find out which detection events correspond to entangled photon pairs. From these events they extract those cases in which they both had used the same basis setting of the analyzers. Owing to the perfect anticorrelations in these cases, Alice and Bob can build a string of bits (the sifted key) by assigning a "0" to the $+1$ results and a "1" to the -1 result of the individual polarization measurements. In order to obtain identical sets of a random bit sequence, one of them finally has to invert the bits.

3.2.5.2   An Entanglement-Based Quantum
Cryptography Prototype System

We have recently developed a quantum cryptography prototype system in cooperation with the Austrian Research Centers Seibersdorf (ARCS). It was

---

*This is due to the nonvanishing probability of producing more than two photons per faint pulse. One possible attack on the security would then simply involve a beamsplitter, which distributes one photon of a pulse to Eve and one to Bob. This would allow Eve to gain sufficient information to reconstruct the distributed key. True single-photon sources are needed to overcome this sufficiency [68].

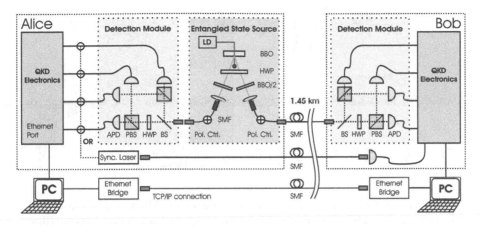

Figure 3.9   Sketch of the experimental setup from [71]. Our entangled state source produces polarization-entangled photon pairs. One of the photons is locally analyzed in Alice's detection module, while the other is sent over a 1.45 km long single-mode optical fiber (SMF) to the remote site (Bob). Polarization measurement is done in one of two bases (0° and 45°) by using a beam splitter (BS) that randomly sends incident photons to one of two polarizing beamsplitters (PBS). One of the PBS is defined for measurement in the 0 basis, and the other in the 45 basis as the half-wave plate (HWP) rotates the polarization by 45°. The final detection of the photons is done in passively quenched silicon avalanche photodiodes (APD). When a photon is detected in one of Alice's four APDs, an optical trigger pulse is created (sync. laser) and sent over a second fiber to Bob to establish a common time basis. At both sites, this trigger pulses and the detection events from the APDs are fed into a dedicated quantum key generation (QKG) device for further processing.

applied in a real-world scenario in April 2004. This was the first time that a quantum cryptography system was used for the encryption of an Internet bank transfer [71]. The system was installed at the headquarters of a large bank (Alice) and at the Vienna City Hall (Bob), and a key was distributed over the 1.45 km optical single-mode fiber connecting the parties.

The quantum cryptography system (Figure 3.9) consists of a portable source for polarization-entangled photons (Figure 3.10) and two sets of fourfold single-photon detection modules with integrated polarization analyzers and embedded hardware devices that are capable of handling the complete software protocol needed to extract a secure and private key out of raw detection events. The quantum channel between Alice and Bob consisted of an optical fiber that has been installed between the two experimental sites in the Vienna sewage system. The classical protocol in that experiment is performed via a standard TCP/IP connection. The exposure of the fibers to realistic environmental conditions, such as stress and strain during installation as well as temperature changes, was an important feature of this experiment; the successful operation of the system shows that laboratory conditions are not necessary for its operation.

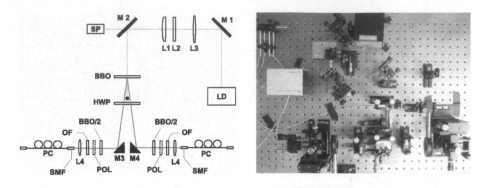

Figure 3.10 Sketch of the experimental implementation. The beam of a laser diode (LD) is focused by a telescope (lenses L1, L2, and L3) onto the nonlinear crystal (BBO). The photon pairs created by SPDC leave the crystal with an opening angle of 6°, passing a half-wave plate (HWP); the polarization of the photons is flipped before they pass the two compensation crystals (BBO/2). Before they are coupled into a single-mode optical fiber (SMF) by the coupling lens L4, they pass an optional polarizer (POL) and an orange glass filter (OF) that blocks scattered UV light. To compensate for arbitrary polarization rotation within the fiber, a polarization control module (PC) is connected to the end of the SMF.

3.2.5.2.1 The Photon Source. The entangled-photon source has been inspired by the design of several previous experiments [72,73] (see Figure 3.10). There, the entangled photons are created by spontaneous parametric down-conversion in a nonlinear crystal. In our setup, a continuous wave (CW) violet GaN laser diode was used to pump a nonlinear β-barium borate (BBO) crystal with 18.6 mW optical power at 405 nm. To achieve a narrow line width, the laser diode was mounted in Littrow configuration.

The pump beam was focused to a round waist of approximately 100 µm at the BBO crystal using a telescope lens system. Three lenses form an imaging system in which an achromatic lens creates a distortion-free elliptical focus that could then be imaged, astigmatically corrected, into the crystal. The Rayleigh range of the pump beam is much longer than the length of the BBO crystal used (4 mm long). We used a half-wave plate and a 2 mm long BBO crystal for compensating for the transversal and longitudinal walkoff effects [72] (see Figure 3.10). We assumed a gaussian distribution of angles and rotational symmetry around the intersection lines for the emission of entangled photons [73]. The collection efficiency was optimized by matching the emission modes of the entangled photons with the modes accepted by the fiber coupler.

The setup is aligned to produce the maximally entangled Bell singlet state

$$|\psi^-\rangle_{12} = \frac{1}{\sqrt{2}} \left(|H\rangle_1 |V\rangle_2 - |V\rangle_1 |H\rangle_2 \right). \tag{3.2}$$

To characterize realistically the quality of our source without completely re-constructing the density matrix, we make an assumption on the noise present in our produced output state. Assuming random (white) noise, our produced state becomes

$$\rho_{12} = v|\psi^-\rangle\langle\psi^-|_{12} + \frac{(1-v)}{4}\mathcal{I}_{12}. \tag{3.3}$$

In this model, the quality of our source is entirely described by the two-photon visibility v and the pair production rate per second. The overall number of detected photon pairs was approximately 25,000 pairs per second, and the average visibility was better than $v = 0.95$.

3.2.5.2.2   QKD Electronics.

The prototype of the dedicated quantum key distribution (QKD) hardware currently under development consists of three main computational components: acquisition of the raw key, generation of synchronization pulses, and QKD protocol tasks. All three units are situated on a single printed circuit board. The developed detection logic is implemented in a FPGA and runs at a sampling frequency of 800 MHz, while employing a time window of 10 ns for matching the detection events and synchronization signals.

The board handles the synchronization channel and generates a strong laser pulse whenever a photon counting event is detected at Alice's site. This is ensured by a logical OR connection of the detector channels. The synchronization laser pulse at the wavelength of 1550 nm was sent over a separate single-mode fiber.

A full scale QKD protocol was implemented very recently including data acquisition, error estimation, error correction, implementing the algorithm CASCADE [74], privacy amplification, and a protocol authentication algorithm that ensures the integrity of the quantum channel by using a Töplitz matrix approach. Furthermore, the encryption library modules applied include one-time pad and AES encryption schemes, the latter allowing key exchange on a scale determined by the user.

3.2.5.2.3   Results.

The average total quantum bit error rate (QBER) of the raw key was found to be less than 8% for more than the entire run time of the experiment. An analysis of the different contributions to the QBER showed that about 2.6% originate in imperfections of the detection modules and 1.2% are due to reduced visibility of the entangled state. The rest of the QBER was attributed to the error produced by the quantum channel. The average raw key bit rate in our system was found to be about 80 bits/s after error correction and privacy amplification. This value is mainly limited by the attenuation on the quantum channel, by the detection efficiency of the avalanche photodiodes, and by the electronics.

To conclude, polarization-entangled photon systems provide an excellent alternative for systems based on weak coherent pulses. Our results suggest that the development of a commercial entanglement based quantum cryptography system is not far away.

3.2.6   Toward a Global Quantum Communication Network

3.2.6.1   Free-Space Distribution of Quantum Entanglement

In more recent years, work has begun on extending the reach of quantum communication to longer and longer distances — after all, what good is a quantum phone if you can only call across the room? Clearly, optical photons are the ideal system for quantum communication over distances, owing to their weak interaction with the environment (i.e., long decoherence times) and high speed. The two methods for sharing photons over long distances are through optical fibers or via free-space optical links. Previously, entangled photons have been shared over long distances only in optical fiber up to 50 km [75]. Similar systems were used to perform a Bell inequality experiment that closed the locality loophole [76]. Free-space optical links provide an exciting alternative quantum channel when there is a direct line of sight between two communicating parties. They consist of at least two telescopes — a transmitter and a receiver — which are used to send light over large distances through the air. Free-space links have been used in conjunction with faint laser pulses to implement the BB84 quantum cryptography protocol up to a distance of 23.4 km [77] and even at daylight [78,79,80]. Theoretical studies have shown that quantum communication in optical fiber can be extended to approximately 100 km before attenuation overwhelms the signal [81]. Recent fiber-based experiments already reach this limit. Similar limitations are valid for optical free-space links, which suffer from attenuation in the atmosphere due to aerosols [82] and from atmospheric turbulences, which are eventually limited by the Earth's curvature. Why is this distance of some hundred kilometers not a limit in our optical networks of today? Quantum information suffers from a fragility that is not present in its classical counterpart. For example, classical optical pulses that encode 0's and 1's in an optical network can be detected and regenerated or amplified every so often in repeater stations, effectively extending the range of optical communication indefinitely. However, the polarization state of a single photon cannot be faithfully amplified — this can be seen as a consequence of the no-cloning theorem [83]. This makes the quantum analogue of repeaters much more complicated than their classical counterparts. A quantum repeater [15] is in principle possible with the use of quantum memories, entanglement purification [29,18], and entanglement swapping [84,85]. In addition, free-space optical links may be the way to increase significantly the present quantum communication distance limit: while earthbound free-space links are just as limited as fiber, they have the advantage in that they can be combined with satellites. The atmosphere is relatively thin, and most of the absorption takes place near the Earth's surface. The attenuation experienced on a clear day at the Earth's surface over approximately 4 km is roughly equivalent to that experienced vertically through the atmosphere [86]. Transmitting entangled photons from space to Earth will definitely allow us to overcome the current distance limits and bridge distances much larger than those achievable with purely ground-based laboratories.

3.2.6.1.1 Free-Space Optical Links with Entangled Photons.

Our group recently demonstrated how combine free-space optical technology with entangled photon pairs. The first experiment took place over the Danube in Vienna, where we could demonstrate the distribution of entangled photon pairs over 600 m [87]. The second experiment [88] was set up over the city of Vienna and achieved a distance of approximately 8 km, which exceeds the atmospheric attenuation for satellite communication.

The schematic setup of the Danube experiment is shown in Figure 3.11. The compact, portable down-conversion source (see Section 2.5) was placed

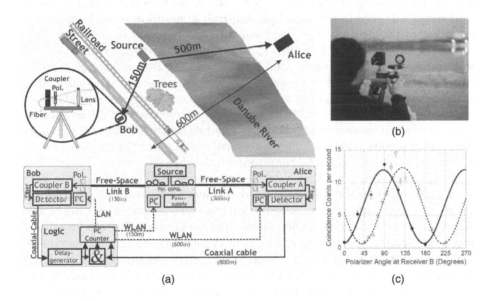

(a)

(b)

(c)

Figure 3.11 (a) Experimental schematic and communication diagram from [87]. The upper figure shows the positioning of the source and receiver stations for the experiment. The down-conversion source was positioned on the southwest bank of the Danube River. One receiver station, named Alice, was located 500 m away on a rooftop on the northeast side of the river, while the second receiver, Bob, was located on a second rooftop 150 m away across a railroad and a highway. The inset shows a schematic of the receiver telescope (the sender is the same with no polarizer). The lower figure shows how data were communicated and shared for the experiment (see text). (b) One of the transmitter telescopes during alignment. (c) Measured polarization correlations between receivers. The data show the measured coincidence rate (per second) between the two receivers as a function of the angle of the polarizer at receiver B when the polarizer at receiver A was set to 0° (solid circles, solid line) and to 45° (open circles, dotted line), respectively. The obviously noisy part in the data coincides with the passing of a freight train underneath the link to receiver B. The visibilities of the best fit curves are 88.2 ± 5.7% and 89.0 ± 3.3%.

on one bank of the Danube and stored in a shipping crate. One receiver station, Alice, was on the far bank of the Danube and located on a rooftop approximately 500 m away. The second receiver, Bob, was located about 150 m from the source location on a second rooftop. Although both receiver stations were located above ground level, the Alice link was periodically blocked by passing ships, and the Bob link, while not completely blocked, experienced extra beam fluctuations from passing freight trains on the railroad. Our diode-laser-pumped down-conversion source requires a fraction of the electrical power and none of the water cooling of an argon-ion- or titanium-sapphire-pumped system. For this experiment, all electrical power for the source was supplied from a gas-powered 2 kW generator. This demonstrates that down-conversion sources are no longer tied to the laboratory environment and can be taken virtually anywhere; they can function in real-world applications. In addition to using the diode-pumped system, we took advantage of a second recent advance in entangled photon-pair generation — high efficiency coupling of the down-conversion light into single-mode optical fiber [73]. Transmission from and collection into nighttime ambient background sources without resorting to high-loss band-pass filters. The transmitting telescopes for the experiment were simply single-mode fiber couplers and a 5-cm achromatic lens with a 150 mm focal length. The receiver telescopes were identical except for a polarizer placed in front of the coupler that could be rotated for polarization measurements. Our singles rate background level was limited to about 600–700 Hz, which was essentially due only to the dark counting rates of the detectors. The communication schematic for the experiment is shown in Figure 3.11. Alice's detection signals were sent directly to Bob via a long coaxial cable that connected the two labs. At the Bob station, a delay generator was used to account for the extra propagation time of the Alice signal and synchronize the coincident pulses, which were measured using standard NIM electronics. While the singles rates and coincidence rates were measured only at the Bob station, the results were distributed via local area network (LAN) connections to the Bob rooftop and Wave-LAN to the source and Alice station. This allowed for remote polarization compensation, telescope adjustments, and data accumulation using only a single measurement configuration. The source parameters have already been described in detail in Section 3.2.5. In short, the polarization-entangled singlet Bell state $|\Psi^-\rangle_{12} = \frac{1}{\sqrt{2}}(|HV\rangle_{12} - |VH\rangle_{12})$ could be generated with a two-photon visibility of approximately $v = 0.95$ with singles rates and coincidence rates of approximately 120,000 Hz and 20,000 Hz, respectively, at a UV pumping power of 18 mW. The light was coupled through the optical telescope links, each of which had an attenuation of 12 dB (or about 6% transmission). This was sufficient to yield singles count rates at the receivers of about 4000 s^{-1} (including background) and a maximum coincidence rate of 15 s^{-1}.

In order to support our claim that the shared photons were entangled, we measured a set of polarization correlations designed to violate maximally a CHSH Bell inequality [89,90] for the singlet. We define a polarization correlation as

$$E(\phi_A, \phi_B) = \frac{N^{++} + N^{--} - N^{+-} - N^{-+}}{N^{++} + N^{--} + N^{+-} + N^{-+}}, \qquad (3.4)$$

where N is the number of coincidence counts when the polarizer is set to the angle ϕ_A ("+") or $\phi_A^{\perp}$ ("−") at Alice and when the polarizer is set to the angle ϕ_B ("+") or $\phi_B^{\perp}$ ("−") at Bob. The CHSH Bell inequality, which holds for any local realistic description of the photon pair's polarization states, is then written as a combination of such polarization correlations for a set of angles; this inequality is

$$S = |E(\phi_A, \phi_B) - E(\phi_A, \tilde{\phi}_B) + E(\tilde{\phi}_A, \phi_B) + E(\tilde{\phi}_A, \tilde{\phi}_B)| \leq 2, \qquad (3.5)$$

where S is the so-called Bell parameter and $\phi_{A(B)}$ and $\tilde{\phi}_{A(B)}$ represent different polarization settings for Alice (Bob). For a pure singlet state, quantum mechanics predicts a maximal violation of this inequality of $S = 2\sqrt{2}$, for the set of angles $\{\phi_A, \tilde{\phi}_A, \phi_B, \tilde{\phi}_B\} = \{0°, 45°, 22.5°, 67.5°\}$.

The experimentally obtained polarization correlations are $E(0°, 22.5°) = -0.509 \pm 0.057$, $E(0°, 67.5°) = +0.643 \pm 0.042$, $E(45°, 22.5°) = -0.558 \pm 0.055$, and $E(45°, 67.5°) = -0.702 \pm 0.046$. Using these results, we calculate $S_{EXP} = 2.41 \pm 0.10$, which is a sufficient violation of the Bell inequality by over four standard deviations. It is also the experimental signature of shared entangled states between the two receiver stations. This work was the first demonstration of the distribution of entangled photon pairs over free-space optical links. A cryptographic system based on our setup would have shown a total raw key generation rate of a few tens of bits per second and an estimated quantum bit error rate (QBER) of 8.4%. It is interesting to note that our link attenuation of 12 dB corresponds to a value that might be achievable with state-of-the-art space technology when establishing a free-space optical link between an Earth-based receiver telescope of 100-cm diameter and a satellite-based transmitter telescope of 20-cm diameter orbiting Earth at a distance of 600 km [91]. Typical losses in an actual satellite experiment might vary, depending on the link optics and on the performance of satellite pointing and tracking [92,93].

In an extended experiment, we could significantly increase the distance between the stations. We have distributed entangled photons between an old observatory and a modern office skyscraper in Vienna, that are 7.8 km apart [88]; see Figure 3.12. The source of the entangled photons is placed at the observatory. The reason for choosing such a distance is that in a 4.5-km link along the ground, one expects the same level of attenuation from scattering with airborne particles as in going through the whole atmosphere vertically.* In order to have a reasonable signal at this distance, we have built redesigned

---

*The transmission of 800 nm light from the whole vertical atmosphere is about 80% under good weather conditions [94,93]. The horizontal attenuation coefficient measured in Vienna was approximately $\alpha = 0.05$ km^{-1}. The horizontal distance with the same attenuation as the whole atmosphere vertically is $L = -ln(0.8)/\alpha = 4.5$ km [82].

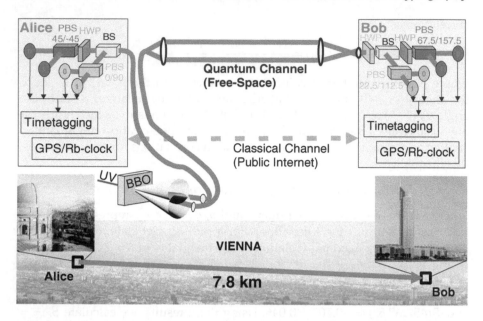

Figure 3.12 Scheme of the free-space quantum communication experiment over Vienna. The transmitter Alice, comprising the single-mode fiber coupled polarization-entangled photon source (DC) and sending telescope, is located in the 19th-century observatory Kuffner Sternwarte. Bob has a receiver telescope and is located on the 46th floor of the Millennium Tower skyscraper 7.8 km away. Alice measures the photons in mode A from each entangled pair using a four-channel detector made of a 50/50 beam splitter (BS), a half-wave plate (HWP), and polarizing beam splitters (PBS), which measures the photon polarization on either the H/V or $+/-$ basis, where $\pm = \frac{1}{\sqrt{2}}(H \pm V)$. She sends the other photon in mode B, after polarization compensation (Pol.), via her telescope and free-space link to Bob. Bob's receiver telescope is equipped with a similar four-channel detector and can measure the polarizations in the same bases as Alice or, by rotating an extra HWP, measure another pair of complementary linear polarization bases. Alice and Bob are both equipped with time-tagging cards, which record the times at which each detection event occurs. Rubidium atomic clocks provide good relative timing stability between the local measurements. Both stations also embed a 1 pps signal from the global positioning system (GPS) into their time-tag data stream to give a well-defined zero time offset. During accumulation, Bob transmits his time tags in blocks over a public Internet channel to Alice. She finds the coincident photon pairs in real time by maximizing the cross-correlation of these time tags. Which of the four detector channels fired is also part of each time tag and allows Alice and Bob to determine the polarization correlations between their coincident pairs. Alice uses her polarization compensators to establish singlet-like anticorrelations between her measurements and Bob's.

refractor telescopes. Our new designs are based on larger and higher quality optical elements. We also relaxed our spatial filtering requirement to reduce sensitivity to beam wander and fluctuations. Using locally recorded time stamps and a public Internet channel, coincident counts from correlated photons are demonstrated to violate a Bell inequality by 14 standard deviations. This confirms the high quality of the shared entanglement and it is an encouraging step toward satellite-based distribution of quantum entanglement and future intracity quantum networks.

3.2.6.2   Quantum Communications in Space

Although free-space optical links are in general superior to optical fibers with respect to photon absorption, terrestrial free-space links will eventually suffer from obstruction of objects in the line of sight, from possible severe attenuation due to weather conditions and aerosols [82] from atmospheric turbulence, and from the Earth's curvature. They are thus limited to rather short distances. To exploit fully the advantages of free-space links, it will be necessary to use space and satellite technology. By transmitting and/or receiving either photons or entangled photon pairs to and/or from a satellite, entanglement can be distributed over truly large distances and thus would allow quantum communication applications on a global scale. Such a scenario looks unrealistic at first sight, but we have recently shown that demonstrations of quantum communication protocols using satellites are already feasible today [91, 96, 101].

Based on present-day technology and assuming reasonable link parameters, one can achieve enough entangled photons per receiver pair to demonstrate several quantum communication protocols. For example, a single optical link between a satellite based transmitter terminal and an optical ground station would suffice to establish a (single-photon) quantum cryptography protocol such as BB84 and hence to generate a secure key between the satellite and the ground station. If the same terminal generates another key with another ground station (at an arbitrary distance from the first one), classical communication between the two ground stations suffices to establish a secret key between them. In other words, satellite-based single-photon links already allow quantum key distribution on a global scale. Note, however, that in this scenario the security requirements on the satellite are as high as for standard cryptography schemes. In contrast, these requirements are relaxed if one can fully exploit an entangled source that distributes pairs of entangled photons to two ground stations. For example, assuming a LEO based transmitter terminal, a simultaneous link to two separate receiving ground stations (see Figure 3.13) and a (conservatively estimated) total link attenuation of approximately 51 dB, one can expect a local count rate of approximately 2600 per second in total at each of the receiver terminals. The number of shared entangled photon pairs is then expected to be approximately 4 per second. For a link duration of 300 seconds, this accumulates to a net reception of 1200 entangled qubits. One can expect erroneous detection events on the order of 7 per 100 seconds, which yields a bit error of approximately 2%. This would already allow a quantum key distribution protocol between the two receiver stations. It is thus clear that a demonstration of basic quantum communication protocols based on quantum entanglement can already be achieved today. Furthermore, the possibility of distributing entangled particles over

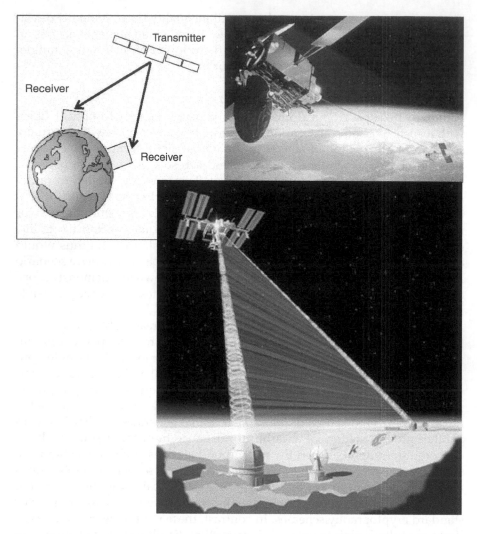

Figure 3.13 Quantum entanglement for space experiments (space-QUEST). Scheme for satellite-based distribution of entangled photons (left: schematic (from [96]); lower right: simulation for source on ISS and two specific ground stations (courtesy of ESA General Studies, Copyright ESA-Autigravite)). Laser comunication satellite terminals such as SILEX (upper right; courtesy of ESA) might provide the technology necessary to establish the optical links between satellites or between satellite and ground stations.

distances beyond the capabilities of earthbound laboratories provides novel opportunities for fundamental tests of quantum physics [95, 101].

Although one must not underestimate the demanding technological challenges associated with bringing quantum entanglement into space, the next steps are both clear and feasible. They include the development of a next generation of space-proof sources for entangled photons as well as the development of space-based and ground-based transmitter and receiver concepts for quantum communication hardware. On the space-terminal side, we have started to investigate the possibility of incorporating an entangled photon source onboard (existing) laser communication satellite terminals [96]. On the ground station side, we have performed the first proof-of-concept tests to demonstrate the feasibility of adapting existing optical ground stations for satellite-ground quantum communication [97]. It may not be too long until the first space-based quantum communication experiment with entangled photons will take off.

3.3 Conclusion and Outlook

Quantum communication has come the long way from purely fundamental considerations on the nature of quantum physics to the implementation of novel concepts and technologies of information processing. Somewhat as a surprise, the role of quantum entanglement has also become more and more important for the applications. It is now a relevant resource for most advanced quantum communication schemes and even for novel quantum computing architectures such as the one-way quantum computer [98,99]. Main future experimental challenges in the field of quantum communication and quantum computation certainly include the development of more reliable and more efficient sources for entanglement. For example, space-based experiments will require a compact, robust, and efficient source for entangled pairs, while quantum computing schemes and multiparty quantum communication schemes will benefit from high-fidelity sources of multiparticle entangled states. Also, improving the interface between photonic qubits and stationary systems such as atoms or solids will eventually allow the realization of quantum memories, which are the major building blocks in any quantum communication network.

We have presented a collection of recent advances in the field of entanglement-based quantum communication. It is fascinating to observe how the last years have paved the way for key quantum technologies such as a quantum repeater and even the realization of a satellite-based global quantum communication network. The quantum physics community is about to reach a stage in which the developed concepts and techniques of quantum communication, which started from curiosity about fundamental aspects of the nature of quantum physics, will evolve into technologies and commercial products on an industrial level. Quantum cryptography has already reached this stage. We are confident that other technologies will soon follow.

Acknowledgments

We acknowledge discussions and collaborations with our scientific colleagues and friends. Our work was supported by the Austrian Science Foundation under Project SFB1506; the European Commission under projects IST-QuComm, RamboQ and SECOQC; the Austrian Research Center Seibersdorf; the

Austrian and European Space Agencies ASA and ESA; and the city of Vienna. We also acknowledge support from Wien Kanal AG, the Austrian BA-CA Bank, Energie AG Oberosterreich, Stumpf AG, Wienerberger AG, the Alexander von Humboldt Foundation (M.A.), the NSERC (K.R.), and the Marie-Curie Program of the EC (G.M.-T.).

References

1. D. Bouwmeester, A. Ekert, and A. Zeilinger, *The Physics of Quantum Information*, Springer-Verlag, Berlin, 2000.
2. M.A. Nielsen and I.L. Chuang, *Quantum Computation and Quantum Information*, Cambridge University Press, Cambridge, U.K., 2000.
3. E. Schrödinger, Die gegenwärtige Situation in der Quantenmechanik, *Naturwissenschaften*, 23, 807–812; 823–828; 844–849, 1935.
4. C.H. Bennett and G. Brassard, Quantum cryptography: public key distribution and coin-tossing, *Proceedings of IEEE International Conference on Computers, Systems and Signal Processing*, Bangalore, India, December 1984, pp. 175–179.
5. C.H. Bennett, F. Bessette, G. Brassard, L. Salvail, and J. Smolin, Experimental quantum cryptography, *J. Crypt.*, 5, 3–28, 1992.
6. C.H. Bennett, G. Brassard, and N.D. Mermin, Quantum cryptography without Bell's theorem, *Phys. Rev. Lett.*, 68, 557–559, 1992.
7. K. Mattle, H. Weinfurter, P.G. Kwiat, and A. Zeilinger, Dense coding in experimental quantum communication, *Phys. Rev. Lett.*, 76(25), 4656–4659, 1996.
8. C.H. Bennett, G. Brassard, C. Crépeau, R. Jozsa, A. Peres, and W.K. Wootters, Teleporting an unknown quantum state via dual classical and Einstein–Podolsky-Rosen channels, *Phys. Rev. Lett.*, 70(13), 1895–1899, 1993.
9. D. Bouwmeester, J.-W. Pan, K. Mattle, M. Eibl, H. Weinfurter, and A. Zeilinger, Experimental quantum teleportation, *Nature*, 390, 575, 1997.
10. N.A. Peters, J.B. Altepeter, D. Branning, E.R. Jeffrey, T.-C. Wei, and P.G. Kwiat, Maximally entangled mixed states: creation and concentration, *Phys. Rev. Lett.*, 92, 133601, 2004.
11. P. Walther, J.-W. Pan, M. Aspelmeyer, R. Ursin, S. Gasparoni, and A. Zeilinger, De Broglie wave of a nonlocal 4-photon, *Nature*, 429, 158–161, 2004.
12. C.H. Bennett, G. Brassard, C. Crépeau, R. Jozsa, A, Peres, and W.K. Wootters, Teleporting an unknown quantum state via dual classical and Einstein-Podolsky-Rosen channels, *Phys. Rev. Lett.*, 70(13), 1895–1899, 1993.
13. D. Gottesman and I.L. Chuang, Demonstrating the viability of universal quantum computation using teleportation and single-qubit operations, *Nature*, 402, 390–393, 1999.
14. E. Knill, R. Laflamme, and G. Milburn, A scheme for efficient quantum computation with linear optics, *Nature*, 409, 46–52, 2000.
15. H.-J. Briegel, W. Dür, J.I. Cirac, and P. Zoller, Quantum repeaters: the role of imperfect local operations in quantum communication, *Phys. Rev. Lett.*, 81, 5932–5935, 1998.
16. T. Jennewein, G. Weihs, J.-W. Pan, and A. Zeilinger, Experimental nonlocality proof of quantum teleportation and entanglement swapping, *Phys. Rev. Lett.*, 88, 17903, 2002.

17. M. Żukowski, A. Zeilinger, M.A. Horne, and A.K. Ekert, "Event-ready-detectors." Bell experiment via entanglement swapping, *Phys. Rev. Lett.*, 71(26), 4287–4290, 1993.

18. J.-W. Pan, S. Gasparoni, M. Aspelmeyer, T. Jennewein, and A. Zeilinger, Freely propagating teleported qubits, *Nature*, 421, 721, 2003.

19. A. Furusawa, J.L. Sorensen, S.L. Braunstein, C.A. Fuchs, H.J. Kimble, and E.S. Polzik, Unconditional quantum teleportation, *Science*, 282, 706–709, 1998.

20. M.D. Barrett, J. Chiaverini, T. Schaetz, J. Britton, W.M. Itano, J.D. Jost, E. Knill, C. Langer, D. Leibfried, R. Ozeri, and D.J. Wineland, Quantum teleportation with atomic qubits, *Nature*, 429, 737–739, 2004.

21. M. Riebe, H. Häffner, C.F. Roos, W. Hänsel, J. Benhelm, G. P. Lancaster, T. W. Körber, C. Becher, F. Schmidt-Kaler, D. F. V. James, and R. Blatt, Quantum teleportation with atoms, *Nature*, 429, 734–737, 2004.

22. S. Bose, V. Vedral, and P.L. Knight, Multiparticle generalization of entanglement swapping, *Phys. Rev. A*, 57, 822–829, 1998.

23. R. Ursin, T. Jennewein, M. Aspelmeyer, R. Kaltenbaek, M. Lindenthal, P. Walther, and A. Zeilinger, Quantum teleportation across the Danube, *Nature*, 430, 849, 2004.

24. N. Lütkenhaus, J. Calsamiglia, and K.-A. Suominen, Bell measurements for teleportation, *Phys. Rev. A*, 59, 3295–3300, 1999.

25. M. Michler, K. Mattle, H. Weinfurter, and A. Zeilinger, Interferometric Bell-state analysis, *Phys. Rev. A*, 53, R1209–R1212, 1996.

26. S. Popescu, Bell's inequalities and density matrices, revealing "hidden" non-locality, *Phys. Rev. Lett.*, 74, 2619–2622, 1995.

27. C.H. van der Wal, M.D. Eisaman, A. Andre, R.L. Walsworth, D.F. Phillips, A.S. Zibrov, and M.D. Lukin, Atomic memory for correlated photon states, *Science*, 301, 196, 2003.

28. A. Kuzmich, W.P. Bowen, A.D. Boozer, A. Boca, C.W. Chou, L.-M. Duan, and J.H. Kimble, Generation of nonclassical photon pairs for scalable quantum communication with atomic ensembles, *Nature*, 423, 731, 2003.

29. C.H. Bennett, C.A. Fuchs, and J.A. Smolin, Entanglement-enhanced classical communication on a noisy quantum channel, in *Proc. 3d Int. Conf. on Quantum Communication and Measurement*, C.M. Caves, O. Hirota, and A.S. Holevo, eds., Plenum, New York, 1997.

30. J.-W. Pan, S. Gasparoni, R. Ursin, G. Weihs, and A. Zeilinger, Experimental entanglement purification, *Nature*, 423, 417–422, 2003.

31. J.-W. Pan, C. Simon, C. Brukner, and A. Zeilinger, Entanglement purification for quantum communication, *Nature*, 410, 1067, 2001.

32. P.W. Shor, Scheme for reducing decoherence in quantum computer memory, *Phys. Rev. A*, 52, R2493–R2496, 1995.

33. A. Steane, Multiple particle interference and quantum error correction, *Proc. R. Soc. Lond. A*, 452, 2551, 1995.

34. E. Knill and R. Laflamme, A theory of quantum error correcting codes, *Phys. Rev. A*, 55, 500, 1997.

35. A. Steane, The ion trap quantum information processor, *Appl. Phys. B*, 64, 623–642, 1997.

36. H.J. Briegel, R. Raussendorf, and A. Schenzle, Optical lattices as a playground for studying multiparticle entanglement, in *Laserphysics at the Limit*, H. Figger, D. Meschede, and C. Zimmerman, eds., Springer, Heidelberg, 2002, pp. 433–477.

37. E. Knill, R. Laflamme, and G. Milburn, A scheme for efficient quantum computation with linear optics, *Nature*, 409, 46–52, 2000.
38. T.B. Pittman, M. Fitch, B. Jacobs, and J. Franson, Experimental controlled-NOT logic gate for single photons in the coincidence basis, *Phys. Rev. A*, 68, 032316, 2003.
39. J.L. O'Brien, G.J. Pryde, A.G. White, T.C. Ralph, and D. Branning, Demonstration of an all-optical quantum controlled-NOT gate, *Nature*, 426, 264, 2003.
40. K. Sanaka, T. Jennewein, J.-W. Pan, K. Resch, and A. Zeilinger, Experimental nonlinear sign shift for linear optics quantum computation, *Phys. Rev. Lett.*, 92(1), 017902, 2004.
41. S. Gasparoni, J.-W. Pan, P. Walther, T. Rudolph, and A. Zeilinger, Realization of a photonic controlled-NOT gate sufficient for quantum computation, *Phys. Rev. Lett.*, 93(2), 020504, 2004.
42. T.B. Pittman, B. Jacobs, and J. Franson, Demonstration of nondeterministic quantum logic operations using linear optical elements, *Phys. Rev. Lett.*, 88, 257902, 2002.
43. P.G. Kwiat, H. Weinfurter, T. Herzog, A. Zeilinger, and M. Kasevich, Experimental realization of interaction-free measurements, *Ann. N.Y. Acad. Sci.*, 755, 383–393, 1995.
44. M. Zukowski, A. Zeilinger, and H. Weinfurter, Entangling photons radiated by independent pulsed sources, *Ann. N.Y. Acad. Sci.*, 755, 91–102, 1995.
45. P. Walther and A. Zeilinger, Experimental realization of a photonic Bell-state analyzer, *quant-ph/0410244*.
46. H. Bechmann-Pasquinucci and A. Peres, Quantum cryptography with 3-state systems, *Phys. Rev. Lett.*, 85, 3313, 2000.
47. H. Bechmann-Pasquinucci and W. Tittel, Quantum cryptography using larger alphabets, *Phys. Rev. A*, 61, 62308–62313, 2000.
48. N.J. Cerf, M. Bourennane, A. Karlsson, and N. Gisin, Security of quantum key distribution using d-level systems, *Phys. Rev. Lett.*, 88, 0127902, 2002.
49. D. Kaszlikowski, P. Gnacinski, M. Żukowski, W. Miklaszewski, and A. Zeilinger, Violations of local realism by two entangled n-dimensional systems are stronger than for two qubits, *Phys. Rev. Lett.*, 85, 4418, 2000.
50. T. Durt, D. Kaszlikowski, and M. Zukowski, Security of quantum key distributions with entangled qudits, *Phys. Rev. A*, 64, 024101, 2001.
51. D. Collins, N. Gisin, N. Linden, S. Massar, and S. Popescu, Bell inequalities for arbitrarily high-dimensional systems, *Phys. Rev. Lett.*, 88, 040404, 2002.
52. P. Horodecki, Separability criterion and inseparable mixed states with positive partial transposition, *Phys. Lett. A*, 233, 233, 1997.
53. C.H. Bennett, D.P. DiVincenzo, T. Mor, P.W. Shor, J.A. Smolin, and B.M. Terhal, Unextendible product bases and bound entanglement, *Phys. Rev. Lett.*, 82, 5385–5388, 1999.
54. R. Jozsa and J. Schlienz, Distinguishability of states and Von Neumann entropy, *Phys. Rev. A*, 62, 012301, 2000.
55. R.T. Thew, A. Acin, H. Zbinden, and N. Gisin, Experimental realization of entangled qutrits for quantum communication, *quant-ph/0307122*.
56. M. Zukowski, A. Zeilinger, and M.A. Horne, Realizable higher-dimensional two-particle entanglements via multiport beam splitters, *Phys. Rev. A*, 55, 2564, 1997.
57. A. Mair, A. Vaziri, G. Weihs, and A. Zeilinger, Entanglement of the orbital angular momentum states of photons, *Nature*, 412, 313, 2001.

58. G. Molina-Terriza, J.P. Torres, and L. Torner, Management of the angular momentum of light: preparation of photons in multidimensional vector states of angular momentum, *Phys. Rev. Lett.*, 88, 013601, 2002.
59. A. Vaziri, J. Pan, T. Jennewein, G. Weihs, and A. Zeilinger, Concentration of higher dimensional entanglement: qutrits of photon orbital angular momentum, *Phys. Rev. Lett.*, 91, 227902, 2003.
60. A. Vaziri, G. Weihs, and A. Zeilinger, Experimental two-photon, three-dimensional entanglement for quantum communication, *Phys. Rev. Lett.*, 89, 240401, 2002.
61. J.P. Torres, Y. Deyanova, L. Torner, and G. Molina-Terriza, Preparation of engineered two-photon entangled states for multidimensional quantum information, *Phys. Rev. A*, 67, 052313, 2003.
62. A. Vaziri, G. Weihs, and A. Zeilinger, Superpositions of the orbital angular momentum for applications in quantum experiments, *J. Opt. B*, 4, S47–S50, 2002.
63. G. Molina-Terriza, A. Vaziri, J. Rehacek, Z. Hradil, and A. Zeilinger, Triggered qutrits for quantum communication protocols, *quant-ph/0401183 (2004)*.
64. G. Molina-Terriza, A. Vaziri, R. Ursin, and A. Zeilinger, Experimental quantum coin tossing, *Phys. Rev. Lett.*, 94, 040501, 2005.
65. C.H. Bennett and G. Brassard, in *Proceedings of the International Conference on Computer Systems and Signal Processing*, Bangalore, 1984, p. 715.
66. N. Gisin, G. Ribordy, W. Tittel, and H. Zbinden, Quantum cryptography, *Rev. Mod. Phys.*, 74, 145–195, 2002.
67. A.K. Ekert, Quantum cryptography based on Bell's theorem, *Phys. Rev. Lett.*, 67(6), 661, 1991.
68. P. Grangier, B. Sanders, and J. Vuckovic, (Eds.), Focus issue: Single photons on demand, *N. J. Phys.*, Vol. 6, 2004.
69. G. Brassard, N. Lütkenhaus, T. Mor, and B. Sanders, Limitations on practical quantum cryptography, *Phys. Rev. Lett.*, 85, 1330–1333, 2000.
70. I. Marcikic, H. de Riedmatten, W. Tittel, H. Zbinden, and N. Gisin, Long-distance teleportation of qubits at telecommunication wavelengths, *Nature*, 421, 509, 2003.
71. A. Poppe, A. Fedrizzi, T. Lörunser, O. Maurhardt, R. Ursin, H.R. Böhm, M. Peev, M. Suda, C. Kurtsiefer, H. Weinfurter, T. Jennewein, and A. Zeilinger, Practical quantum key distribution with polarization entangled photons, *Optics Express*, 12, 3865–3871, 2004.
72. P.G. Kwiat, K. Mattle, H. Weinfurter, and A. Zeilinger, New high-intensity source of polarization-entangled photons, *Phys. Rev. Lett.*, 75, 4337, 1995.
73. C. Kurtsiefer, M. Oberparleiter, and H. Weinfurter, High-efficiency entangled photon pair collection in type-II parametric fluorescence, *Phys. Rev. A*, 64, 23802, 2001.
74. G. Brassard and L. Salvail, Secret-key reconciliation by public discussion, in *EUROCRYPT*, T. Helleseth, ed., Vol. 765, Springer, New York, 1993, p. 410.
75. I. Marcikic, H. de Riedmatten, W. Tittel, H. Zbinden, M. Legr, and N. Gisin, Distribution of time-bin entangled qubits over 50 km of optical fiber, *Phys. Rev. Lett.*, 93, 180502, 2004.
76. G. Weihs, T. Jennewein, C. Simon, H. Weinfurter, and A. Zeilinger, Violation of Bell's inequality under strict Einstein locality conditions, *Phys. Rev. Lett.*, 81, 5039–5043, 1998.
77. C. Kurtsiefer, P. Zarda, M. Halder, H. Weinfurter, P. Gorman, P. Tapster, and J. Rarity, A step towards global key distribution, *Nature*, 419, 450, 2002.

78. W.T. Buttler, R.J. Hughes, P.G. Kwiat, S.K. Lamoreaux, G.G. Luther, G.L. Morgan, J.E. Nordholt, C.G. Peterson, and C.M. Simmons, *Phys. Rev. Lett.*, 81, 3283–3286, 1998.

79. W.T. Buttler, R.J. Hughes, S.K. Lamoreaux, G.L. Morgan, J.E. Nordholt, and C.G. Peterson, Daylight quantum key distribution over 1.6 km, *Phys. Rev. Lett.*, 84, 5652–5655, 2000.

80. R.J. Hughes, J.E. Nordholt, D. Derkacs, and C.G. Peterson, Practical free-space quantum key distribution over 10 km in daylight and at night, *New J. Phys.*, 4, 43.1–43.14, 2002.

81. E. Waks, A. Zeevi, and Y. Yamamoto, Security of quantum key distribution with entangled photons against individual attacks, *Phys. Rev. A*, 65, 052310, 2002.

82. H. Horvath, L.A. Arboledas, F.J. Olmo, O. Jovanović, N. Gangl, W. Kaller, C. Sánchez, H. Sauerzopf, and S. Seidl, Optical characteristics of the aerosol in Spain and Austria and its effect on radiative forcing, *J. Geophys. Res.*, 107(D19), 4386, 2002.

83. W.K. Wootters and W.H. Zurek, A single quantum cannot be cloned, *Nature*, 229, 802–803, 1982.

84. M. Zukowski, A. Zeilinger, M.A. Horne, and A.K. Ekert, event-ready-detectors" Bell experiment via entanglement swapping, *Phys. Rev. Lett.*, 71, 4287–4290, 1993.

85. J.-W. Pan, D. Bouwmeester, H. Weinfurter, and A. Zeilinger, Experimental entanglement swapping: entangling photons that never interacted," *Phys. Rev. Lett.*, 80, 3891, 1998.

86. H. Horwath, L.A. Arboledas, F. Olmo, O. Jovanovic, M. Gangl, W. Kaller, C. Sanchez, H. Sauerzopf, and S. Seidl, Optical characteristics of the aerosol in Spain and Austria and its effect on radiative forcing. *J. Geophys. Res. (Atmospheres)*, 107, No. D19, AAC 9, 2002.

87. M. Aspelmeyer, H.R. Böhm, T. Gyatso, T. Jennewein, R. Kaltenbaek, M. Lindenthal, G. Molina-Terriza, A. Poppe, K. Resch, M. Taraba, R. Ursin, P. Walther, and A. Zeilinger, Long-distance free-space distribution of quantum entanglement, *Science*, 301, 621–623, 2003.

88. K.J. Resch, M. Lindenthal, B. Blauensteiner, H.R. Böhm, A. Fedrizzi, C. Kurtsiefer, A. Poppe, T. Schmitt-Manderbach, M. Taraba, R. Ursin, P. Walther, H. Weier, H. Weinfurter, and A. Zeilinger, Distributing entanglement and single photons through an intra-city, free-space quantum channel, *Opt. Express*, 13, 202, 2005.

89. J.S. Bell, On the Einstein Podolsky Rosen paradox, *Physics*, 1, 195–200, 1964.

90. J.F. Clauser, M.A. Horne, A. Shimony, and R.A. Holt, Proposed experiment to test local hidden-variable theories, *Phys. Rev. Lett.*, 23, 880, 1969.

91. M. Aspelmeyer, T. Jennewein, M. Pfenningbauer, W. Leeb, and A. Zeilinger, Long-distance quantum communication with entangled photons using satellites, *IEEE J. Selected Top. Quantum Electron.*, 9, 1541–1551, 2003.

92. R. Hughes, W. Buttler, P. Kwiat, S. Lamoreaux, G. Morgan, J. Nordholt, and C. Peterson, Free-space quantum key distribution in daylight, *J. Mod. Opt.*, 47, 549–562, 2000.

93. J.G. Rarity, P.R. Tapster, P.M. Gorman, and P. Knight, Ground to satellite secure key exchange using quantum cryptography, *New J. Phy.*, 4, 82, 2002.

94. J.E. Nordholt, R. Hughes, G.L. Morgan, C.G. Peterson, and C.C. Wipf, Present and future free-space quantum key distribution, in *Free-Space Laser Communication Technologies XIV*, *Proc. SPIE*, 4635. SPIE, 2002, p. 116.

95. R. Kaltenbaek, M. Aspelmeyer, T. Jennewein, C. Brukner, M. Pfennigbauer, W. Leeb, and A. Zeilinger, Proof-of-concept experiments for quantum physics in space, Proc. of SPIE, *Quantum Communications and Quantum Imaging*, 5161, 252–268, 2003.

96. M. Pfennigbauer, W. Leeb, G. Neckamm, M. Aspelmeyer, T. Jennewein, F. Tiefenbacher, A. Zeilinger, G. Baister, K. Kudielka, T. Dreischer, and H. Weinfurter, Accommodation of a quantum communication transceiver in an optical terminal (ACCOM), Report prepared for the European Space Agency (ESA) under ESTEC/Contract No. 17766/03/NL/PM, 2005.

97. P. Villoresi, F. Tamburini, M. Aspelmeyer, T. Jennewein, R. Ursin, C. Pernechele, G. Bianco, A. Zeilinger, and C. Barbieri, Space-to-ground quantum-communication using an optical ground station: a feasibility study, in *Quantum Communications and Quantum Imaging II*, R. Meyers and V. Shih (eds.), Proc. of SPIE, Vol. 5551, 113–120, 2004.

98. R. Raussendorf and H.J. Briegel, A one-way quantum computer, *Phys. Rev. Lett.*, 86(22), 5188–5191, 2001.

99. P. Walther, K.J. Resch, T. Rudolph, E. Schenck, H. Weinfurter, V. Vedral, M. Aspelmeyer, and A. Zeilinger, Experimental one-way quantum computing, *Nature*, 434, 169, 2005.

100. D.M. Greenberger, M.A. Horne and A. Zeilinger, Going beyond Bell's theorem, in M. Kafatos (ed.), *Bell's theorem, quantum theory, and conceptions of the universe*, Kluwer, Dordrecht, 1989.

101. M. Aspelmeyer, T. Jennewein, H.R. Böhm, C. Brukner, R. Kaltenbaek, M. Lindenthal, G. Molina-Terriza, J. Petschinka, R. Ursin, P. Walther, A. Zeilinger, M. Pfennigbauer and W. Leeb, QSpace — Quantum communications in space, Report prepared for the European Space Agency (ESA) under ESTEC/Contract No. 16358/02/NL/SFe, 2003.

chapter 4

The DARPA Quantum Network

C. Elliott
BBN Technologies

Contents

4.1   Introduction... 83
4.2   Current Status of the DARPA Quantum Network 84
4.3   Motivation for the DARPA Quantum Network........................ 86
4.4   What Is a QKD Network?... 86
     4.4.1   Photonic Switching for "Untrusted Networks" 89
     4.4.2   Key Relay for "Trusted Networks" 89
     4.4.3   The Major Benefits of QKD Networks....................... 89
4.5   BBN's "Mark 2" Weak Coherent Systems 90
4.6   BBN QKD Protocols... 96
4.7   Photonic Switching for Untrusted Networks 97
4.8   BBN Key Relay Protocols for Trusted Networks................... 99
4.9   Future Plans.. 100
4.10   Summary.. 100
Acknowledgments 101
References 101

4.1   Introduction

It now seems likely that quantum key distribution (QKD) techniques can provide practical building blocks for highly secure networks and in fact may offer valuable cryptographic services, such as unbounded secrecy lifetimes, which can be difficult to achieve by other techniques. Unfortunately, however, QKD's impressive claims for information assurance have been to date at least partly offset by a variety of limitations. For example, traditional QKD is distance limited, can only be used across a single physical channel (e.g., free-space or telecommunications fiber, but not both in series due to frequency propagation and modulation issues), and is vulnerable to disruptions such as fiber cuts because it relies on single points of failure.

To a surprising extent, however, these limitations can be mitigated or even completely removed by building QKD *networks* instead of the traditional stand-alone QKD links. Accordingly, a team of participants from BBN Technologies, Boston University, and Harvard University has recently built and begun to operate the world's first quantum key distribution network under Defense Advanced Research Projects Agency (DARPA) sponsorship.*

The DARPA Quantum Network became fully operational on October 23, 2003, in BBN's laboratories, and in June 2004 it was fielded through dark fiber underneath the streets of Cambridge, Massachusetts, to link our campuses with nonstop quantum cryptography, 24 hours per day. It is the world's first quantum cryptography network and indeed probably the first metro-area QKD deployment in continuous operation. As of December 2004, it consists of six QKD nodes. Four are used in BBN-built, interoperable weak-coherent QKD systems running at a 5-MHz pulse rate through telecommunications fiber and inter-connected via a photonic switch. Two are electronics built by the National Institute of Standards and Technology (NIST) for a high-speed free-space QKD system. All run BBN's full suite of production-quality QKD protocols. In the near future, we plan to add four more quantum cryptographic nodes based on a variety of physical phenomena and start testing the resulting network against sophisticated attacks.

This chapter introduces the DARPA Quantum Network as it currently exists and briefly outlines our plans for the near future. We first describe the motivation for our work and define the basic principles of a quantum cryptographic *network* (which may be composed of a number of QKD systems with relays and/or photonic switches). We then discuss the specifics of our current weak-coherent QKD network, including its QKD links, photonic switches for "untrusted" networks, and key relay protocols for "trusted" networks. We conclude with future plans and our acknowledgments.

4.2 Current Status of the DARPA Quantum Network

Figure 4.1 displays a fiber diagram of the DARPA Quantum Network's build-out through Cambridge, Massachusetts, as of December 2004. The network consists of two weak-coherent BB84 transmitters (Alice, Anna), two compatible receivers (Bob, Boris), and a 2 × 2 switch that can couple any transmitter to any receiver under program control. Alice, Bob, and the switch are in BBN's laboratory; Anna is at Harvard; and Boris is at Boston University (BU). The fiber strands linking Alice, Bob, and the switch are several meters long. The

---

*The opinions expressed in this article are those of the author alone and do not necessarily reflect the views of the United States Department of Defense, DARPA, or the United States Air Force.

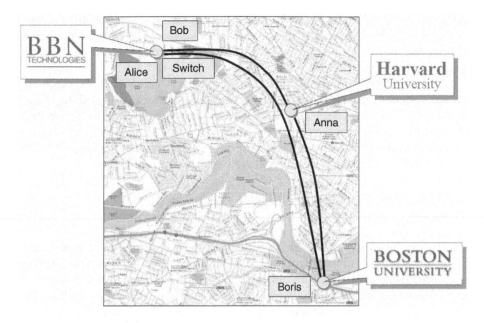

Figure 4.1   The metro-fiber portions of the DARPA Quantum Network.

Harvard-BBN strand is approximately 10 km. The BU-BBN strand is approximately 19 km. Thus the Harvard-BU path, through the switch at BBN, is approximately 29 km. All strands are standard SMF-28 telecommunications fiber. Figure 4.2 presents the network in schematic form.

Anna's mean photon number is 0.5 at present, with the Anna–Bob link delivering about 1,000 privacy-amplified secret bits/second at an average 3% quantum bit error rate (QBER). At present, the DARPA Quantum Network cannot support fractional mean photon numbers to Boris at BU, owing to high attenuation in fiber segments across the Boston University campus and relatively inefficient detectors in Boris. (BBN–BU attenuation is approximately 11.5 dB). Thus the network currently operates at a mean photon number of 1.0

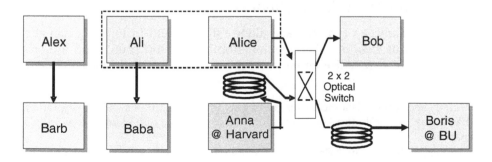

Figure 4.2   Connectivity schematic of the DARPA Quantum Network.

on the BBN–BU link, in order continuously to exercise all parts of the system, even though the resultant secret key yield is zero. In the near future, fiber splices and perhaps detector upgrades should allow operation to BU with mean photon numbers of 0.5.

The DARPA Quantum Network also contains Ali and Baba, the electronics subsystems for a high-speed free-space QKD system designed and built by the NIST. Ali and Baba run the BBN QKD protocols and are linked into the overall network by key relay between Ali and Alice. It further contains two new entanglement based nodes named Alex and Barb, built jointly by BU and BBN, but these nodes are not yet fully operational.

4.3   Motivation for the DARPA Quantum Network

QKD provides a technique for two distinct devices to come to agreement upon a shared random sequence of classical bits, with a very low probability that other devices (eavesdroppers) will be able to make successful inferences as to those bits' values. Such sequences may then be used as secret keys for encoding and decoding messages between the two devices. In short, it is a cryptographic key distribution technique.* Although QKD is an interesting and potentially quite useful technique for key distribution, it is not the only one — human couriers and algorithmic "one-way" functions such as the Diffie–Hellman come immediately to mind — and thus it is important to gauge QKD's strengths across a number of goals for key distribution systems in general. Table 4.1 provides such an assessment of "classic" QKD techniques; see [1] for a more extended treatment of this subject.

It can be seen that "classic" QKD, i.e., QKD performed by sending a single quantum entity directly from source to destination, has areas of weakness mixed with its strengths. As important guidelines of our overall research agenda, we are working to strengthen QKD's performance in these weaker areas. A surprising number of these weaknesses, as it turns out, can be removed by weaving individual QKD links into an overall QKD network such as the DARPA Quantum Network.

4.4   What Is a QKD Network?

Figure 4.3 depicts a typical stand-alone QKD system in highly schematic form.** In this example, Alice contains both a photon source and a modulator; in this case, Alice employs an attenuated laser and Mach–Zehnder

---

*Strictly speaking, it is a means for coming to agreement upon a shared key, rather than a way to distribute a key, but we follow conventional QKD terminology in this chapter.

**In fact, it depicts our "Mark 2" weak-coherent link but corresponding forms of high-level schematics can be drawn for any quantum cryptographic system.

Table 4.1   Assessment of "Classic" Quantum Cryptography

Important Goals for a Cryptographic Key Distribution System	QKD Strengths and Weaknesses
Protection of Keys	QKD offers significant advantages in this regard and indeed this is the main reason for interest in QKD. Assuming that QKD techniques are properly embedded into an overall secure system, they can provide automatic distribution of keys that may offer security superior to that of competitors.
Authentication	QKD does not in itself provide authentication. Current strategies for authentication in QKD systems include prepositioning of secret keys at the distant device, to be used in hash-based authentication schemes, or hybrid QKD–public key techniques.
Robustness	This critical property has not traditionally been taken into account by the QKD community. Since keying material is essential for secure communications, it is extremely important that the flow of keying material not be disrupted, whether by accident or by the deliberate acts of an adversary (i.e., by denial of service). Here QKD has provided a highly fragile service to date, since QKD techniques have implicitly been employed along a single point-to-point link.
Distance and Location Independence	This feature is notably lacking in QKD, which requires the two entities to have a direct and unencumbered path for photons between them, and which can only operate for a few tens of kilometers through fiber.
Resistance to Traffic Analysis	QKD in general has had a rather weak approach, since most setups have assumed dedicated, point-to-point QKD links between communicating entities, which has thus clearly laid out a map of the underlying key distribution relationships.

interferometer. Bob contains another modulator and photon detectors, specifically a twin Mach–Zehnder interferometer and cooled InGaAs APDs. Here the channel between Alice and Bob is a standard telecommunications fiber.

In the network context, this system can be viewed as a single, isolated QKD "link" that allows Alice and Bob to agree upon shared cryptographic key material. In our terminology, both Alice and Bob are QKD endpoints, as are other cryptographic stations based on QKD technology, e.g., the Alice

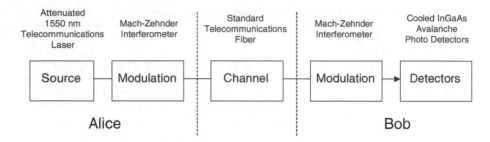

Figure 4.3   High-level schematic of an exemplary QKD link.

or Bob used in plug-and-play systems, entanglement-based links with Ekert protocols, and so forth.

Figure 4.4 shows how a number of such QKD endpoints and links may be woven together into an overall QKD *network*. Here we see that one Alice/Bob pair (A1, B1) is directly connected via a fiber strand, while another (A2, B2) is connected by a free-space channel. Other pairs are connected via fibers and photonic switches, e.g., either A1 or A3 may be connected to B3 depending on the setting of the switch between them. Multiple QKD endpoints, e.g., (B1, A2), may also be grouped into a "key relay device," whose purpose is explained below.

By proper use of QKD networking protocols, a node such as A1 may agree upon key material not just with its direct neighbors (B1 or B3) but indeed with nodes many hops away through the key distribution network. For example, it could agree upon shared keys with B4, through intermediaries of B1, A2, and B4. Perhaps more surprisingly, it could also agree upon shared keys with other transmitters even though neither transmitter can detect the other's photons! Thus two transmitting nodes such as A1 and A3 can agree upon shared keys in a quantum cryptographic network — provided that they rely on a trusted relay, such as B1, to act as a middleman in this process.

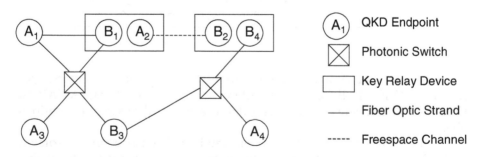

Figure 4.4   High-level schematic of an exemplary QKD network.

4.4.1 Photonic Switching for "Untrusted Networks"

Untrusted networks employ unamplified, all-optical paths through the network mesh of fibers, photonic switches, and endpoints. Thus a photon from its source QKD endpoint proceeds, without measurement, from switch to switch across the optical QKD network until it reaches the destination endpoint at which point it is detected. The (A1, B3) path in Figure 4.4 provides an example, though in general a path may transit multiple photonic switches.

Untrusted QKD networks support truly end-to-end key distribution — QKD endpoints need not share any secrets with the key distribution network or its operators. This feature could be extremely important for highly secure networks. Unfortunately, though, untrusted switches cannot extend the geographic reach of a QKD network. In fact, they may significantly reduce it, since each switch adds at least a fractional dB insertion loss along the photonic path. In addition, it will also prove difficult in practice to employ a variety of transmission media within an untrusted network, since a single frequency or modulation technique may not work well along a composite path that includes both fiber and free-space links.

4.4.2 Key Relay for "Trusted Networks"

After a set of QKD nodes have established pairwise agreed-to keys along an end-to-end path between two QKD endpoints — e.g., (A1, A4) in Figure 4.4 — they may employ these key pairs to relay securely a key "hop by hop" from one endpoint to another, being one-time-pad encrypted and decrypted with each pairwise key as it proceeds from one relay to the next. In this approach, the end-to-end key will appear in the clear within the relays' memories proper but will always be encrypted when passing across a link.

Key relays bring important benefits but are not a panacea. They can extend the geographic reach of a network secured by quantum cryptography, since wide-area networks can be created by a series of point-to-point links bridged by active relays. Furthermore, links can employ heterogeneous transmission media, i.e., some may be through fiber while others are free-space. Thus in theory such a network could provide fully global coverage. However, QKD key relays must be *trusted*. Since keying material and — directly or indirectly — message traffic are available in the clear in the relays' memories, these relays must not fall into an adversary's hands. They need to be in physically secured locations and perhaps guarded if the traffic is truly important. In addition, all users in the system must trust the network (and the network's operators) with all keys to their message traffic.

4.4.3 The Major Benefits of QKD Networks

Table 4.2 summarizes the major benefits that QKD networks bring to traditional, stand-alone QKD links.

Table 4.2   Major Benefits of Quantum Cryptographic Networks

Benefit	Discussion
Longer Distances	QKD key relay can easily extend the geographic reach of quantum cryptography. As one example, quantum cryptography could be performed through telecommunications fiber across a distance of 500 km by interposing four relays between the QKD endpoints, with a span of 100 km fiber between each relay node.
Heterogenous Channels	QKD key relay can mediate between links based on different physical principles, e.g., between free-space and fiber links, or even between links based on entanglement and those based on weak laser pulses. This allows one to "stitch together" large networks from links that have been optimized for different criteria.
Greater Robustness	QKD networks lessen the chance that an adversary could disable the key distribution process, whether by active eavesdropping or simply by cutting a fiber. When a given point-to-point QKD link within the network fails — e.g., by fiber cutting or too much eavesdropping or noise — that link may be abandoned and another used instead. Thus QKD networks can be engineered to be resilient even in the face of active eavesdropping, fiber cuts, equipment failures, or other denial-of-service attacks. A QKD network can be engineered with as much redundancy as desired simply by adding more links and relays to the mesh.
Cost Savings	QKD networks can greatly reduce the cost of large-scale interconnectivity of private enclaves by reducing the required $(N \times N - 1)/2$ point-to-point links to as few as N links in the case of a simple star topology for the key distribution network.

4.5   BBN's "Mark 2" Weak Coherent Systems

This section describes the four fiber based QKD systems currently running in the DARPA Quantum Network. All became operational in October 2003; we call these "Mark 2" systems because they replaced our first-generation system, which started continuous operation in December 2002. These links were inspired by a pioneering Los Alamos system [2].

Each Mark 2 link employs a highly attenuated telecommunications laser (hence the term "weak coherent") at 1550.12 nm, phase modulation via unbalanced Mach–Zehnder interferometers, and cooled avalanche photo detectors (APDs). Most Mark 2 electronics are implemented by discrete components such as pulse generators, though it would not be difficult to integrate all electronics onto a small custom board. Figure 4.5 depicts Anna and Boris, our first rack-mounted versions of the Mark 2 hardware, before their deployment into wiring closets at Harvard and Boston University as part of the metro network.

Figure 4.5   BBN's Mark 2 weak-coherent transmitter and receiver (Anna and Boris).

Two aspects of the Mark 2 weak-coherent link are fundamental to its overall operation and performance. It is important to understand these fundamental points, why they occur, and their implications on the overall system.

First, this link is designed to run through telecommunications fiber as widely deployed today. Thus we have chosen to transmit dim pulses in the 1550.12 nm window for maximal distance through this fiber. At present, these dim pulses can be best detected by certain kinds of commercial APDs cooled to approximately −40 degrees Celsius. These cooled detectors form one of the most important bottlenecks in the overall link performance, as they require on the order of 10 μsec to recover between detection events. The overall link has been designed to run at up to a 5-MHz transmit rate but with a dead time circuit to disable the APD after a detection event in order to accommodate this recovery interval and detector after-pulsing.

Second, the Mark 2 weak-coherent link employs an attenuated telecommunications laser as its source of dim ("single photon") pulses. This is certainly the easiest kind of source to build. Such attenuated, weak-coherent sources, however, have been shown to be vulnerable to at least theoretical forms of attack from Eve by Brassard et al. [3]. Such attacks are generally termed photon number splitting (PNS) attacks. For experimental purposes, we sometimes run the so-called Geneva sifting protocol or SARG [4], which may provide protection against such attacks, even in systems employing attenuated laser pulses. In addition, our forthcoming entangled link will employ a completely different type of source, namely, the BU entangled source. (If possible, we may attempt to build an Eve that can actively attack the attenuated pulses used in our Mark 2 weak-coherent link and perform laboratory demonstrations of this heretofore theoretical form of attack.)

Figure 4.6 highlights the major features of our Mark 2 weak-coherent link. As shown, the transmitter at Alice sends data by means of very highly attenuated laser pulses at 1550.12 nm. Each pulse passes through a Mach–Zehnder interferometer at Alice and is randomly modulated to one of four phases, thus encoding both a basis and a value in that photon's self interference. The receiver at Bob contains another Mach–Zehnder interferometer, randomly set to one of two phases in order to select a basis for demodulation. The received single photons pass through Bob's interferometer to strike one of the two cooled detectors and hence to present a received value. Alice also transmits bright pulses at 1550.92 nm, multiplexed over the same fiber, to send timing and framing information to Bob.

Alice provides the clock source for both transmitter and receiver. All clocking in this system ultimately derives from a single trigger supplied from the transmitter suite. The rising edge of this signal triggers a pulse generator whose output is split into two pulses: one drives the 1550.12 nm QKD data laser to create data pulses, and the other drives the 1550.92 nm sync laser through a gate and delay line. The delay line provides a stable time relationship between the data and sync pulses and is chosen so that the sync pulse is transmitted about 20 ns after its associated data pulse.

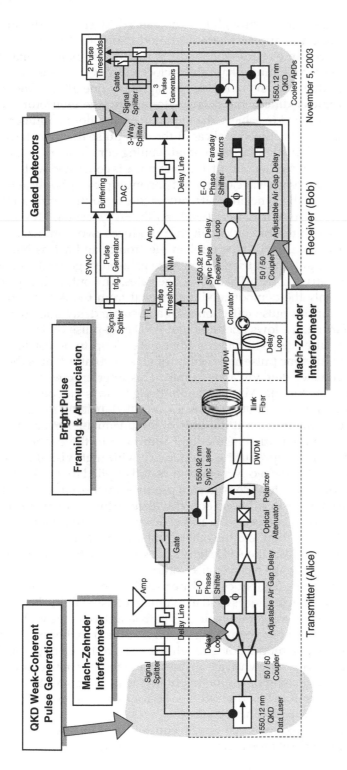

Figure 4.6   Functional decomposition of the Mark 2 weak-coherent QKD link.

The data pulse passes through the unbalanced Mach–Zehnder interferometer where one arm applies phase shift modulation to the pulse. A D-to-A converter drives an electro-optic modulator with an analog voltage that produces the Basis and Value phase shifts clocked from the transmitter electronics. In the other arm an adjustable air gap delay line allows fine-tuning of the interferometer differential delay. After exiting the interferometer the data pulse is attenuated to achieve the requisite mean photon number. A polarizer then removes mistimed replicas of the data pulse that may have been generated by misaligned polarization-maintaining components in the interferometer. At the transmitter output, the data pulse is combined with the sync pulse in a DWDM (dense wavelength division multiplexing) optical filter.

At the receiver the sync and data pulses are separated with a DWDM filter and the sync pulse is detected with a PIN-FET receiver. This signal is shaped in a pulse thresholding circuit that produces two outputs: a 100 ns TTL-level clock signal sent to the receiver electronics and a 4 ns NIM-level APD gate-timing pulse that triggers the APD gate-pulse generators and the pulse generator driving the APD output line gates. The output line gates are timed to pass only the demodulated data signal from the APDs and block noise due to spurious pulse reflections. An adjustable delay line in the NIM pulse interconnection allows fine-tuning of APD gate-pulse timing.

The data pulse passes through a fiber delay loop to adjust its timing with respect to the sync pulse and then through a circulator that is the input to the interferometer demodulation circuit. This interferometer is a folded version of the conventional Mach–Zehnder design and is independent of the input polarization to accommodate the uncontrolled incident polarization at the receiver. Faraday mirrors at the ends of the unequal-length arms reflect light so that the polarization of the light returning to the beam splitter is the same for each arm, producing interference with high visibility [5]. The Basis is clocked out of the receiver electronics and applied to the electro-optic modulator through a D-to-A converter to produce a phase shift of either 0 or $\pi/2$. A pair of cooled APDs, biased above avalanche breakdown only during the time a data photon is expected to arrive, detect the interferometer outputs, one from the beam splitter and the other from the circulator. After gating to select only the data pulse, the APD signals are shaped by threshold detectors and passed as 0 or 1 to the receiver electronics.

A phase-correcting feedback signal, derived by the receiver from training frames sent by the transmitter, is used to maintain phase stability between the transmitter and receiver interferometers as path lengths change with temperature and stress. This phase-correcting signal is applied to the receiver interferometer electrooptic modulator through the transmitter electronics. Phase correction is also necessary when a transmitter and receiver first connect during a startup or switching operation to obtain the phase-matched condition needed for low quantum bit error rate. See [6] for a discussion of BBN's algorithms for automatic path-length control.

It is by now well-known that certain conventional InGaAs APDs can be operated in the single-photon regime if properly cooled and gated. Like many other research teams, we have selected Epitaxx EPM 239 AA APDs for our detectors. Even with this special treatment, they suffer considerably from low quantum efficiency (QE), relatively high dark noise, and serious after-pulsing problems. Even so, they provide adequate performance for a 5-MHz pulse rate quantum cryptography system. Since custom cooling and electronics are required, we designed and built our own cooler package to maintain the InGaAs APDs at the requisite operating temperatures. We had two key goals in mind for the cooler package: it should be able to operate reliably and repeatably over a wide range of temperatures down to –80°C, to enable exploration of detector behavior over a range of operating conditions; and it should be suitable for prototype deployment and thus should not require human intervention on a regular basis, e.g., to refill liquid nitrogen reservoirs.

Figure 4.7 shows a schematic of the core housing, which is a vacuum-pumped chamber containing a pair of InGaAs detectors brought to operating temperature by two Peltier thermoelectric coolers (TECs).

The housing itself is machined aluminum, consisting of a base container plus a removable lid. The lid fits snugly with an O-ring so that a vacuum can be maintained in the inner chamber. A number of holes pass through the base container to allow hermetic feedthrough of fibers and electrical connections. A large connection leads to the vacuum line. At the center of the chamber rests a cooled block of copper with holes drilled out for two detectors. The

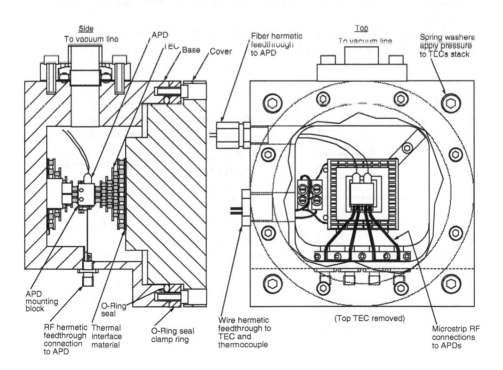

Figure 4.7   Side and top cutaway views of DARPA thermoelectric cooler package.

detectors are inserted into this block, with fibers fed to the outside through a hermetic seal. Microstrip RF connections lead from the side of the container to the detectors. These two detectors are the two 1550.12 nm QKD cooled APDs, i.e., one is D0 and the other is D1. Each detector has its own fiber lead through which come the dim pulses for that detector. Each detector also has a set of microstrip RF connectors by which the bias voltage can be applied and the detector output led to electronics outside the chamber.

The block of copper rests between two five-stage Peltier coolers, also known as thermoelectric coolers (TECs). We chose two coolers of this size to ensure that we had an adequate margin for whatever range of temperatures we wanted to explore; two or three stages would suffice for routine operation. The chamber also contains a thermocouple to measure its current temperature. Electrical leads for the coolers and thermocouple pass through hermetic feedthroughs to the outside equipment.

4.6   BBN QKD Protocols

Although a detailed discussion of BBN's QKD protocols is well beyond the scope of this chapter, quantum cryptographic systems contain a surprising amount of sophisticated software. It has been our observation that the optics in quantum cryptography is perhaps the easiest part; the electronics are more difficult. Moreover, for a real, functional system, the software is harder than the electronics.

Figure 4.8 illustrates our software architecture in a high-level form. Here we see that the QKD protocols have been integrated into a Unix operating

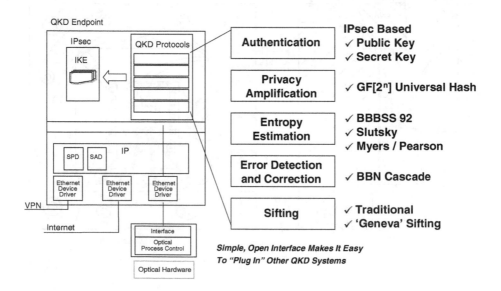

Figure 4.8   The BBN QKD protocol suite in context.

system and provide key material to its indigenous Internet key exchange (IKE) daemon for use in cryptographically protecting Internet traffic via standard IPSec protocols and algorithms. See [7] for a more detailed discussion of this implementation and how QKD interacts with IKE and IPsec.

BBN's QKD protocol stack is an industrial-strength implementation written in the C programming language for ready portability to embedded real-time systems. At present, all protocol control messages are conveyed in IP datagrams so that control traffic can be conveyed via the Internet. However, the control messages could be ported to use other forms of communications quite easily, e.g., ATM networks or dedicated channels.

Two aspects of BBN's QKD protocol stack deserve special mention. First, it implements a complete suite of QKD protocols. In fact, it implements multiple "plug compatible" versions of some functions, as shown in Figure 4.8; for instance, it provides both the traditional sifting protocol and the newer Geneva-style sifting [4].* It also provides a choice of entropy estimation functions. We expect to add additional options and variants as they are developed. Second, BBN's QKD protocols have been carefully designed to make it as easy as possible to plug in other QKD systems, i.e., to facilitate the introduction of QKD links from other research teams into the overall DARPA Quantum Network.

4.7 Photonic Switching for Untrusted Networks

The DARPA Quantum Network currently consists of two transmitters, Alice and Anna, and two compatible receivers, Bob and Boris, interconnected through their key transmission link by a 2 × 2 optical switch. In this configuration, either transmitter can directly negotiate a mutual key with either receiver. The switch must be optically passive so that the quantum state of the photons that encode key bits is not disturbed.

Figure 4.9 depicts the fiber-based portion of the current network diagram. Here all four QKD endpoints are connected through a conventional 2 × 2 optical switch. At one switch position, Alice is connected to Bob, and Anna to Boris. At the other, Alice is connected to Boris, and Anna to Bob. At present the switch controller changes this connectivity on a periodic basis, e.g., every 15 minutes. Immediately after the switch setting is changed, the receivers autonomously discover that they are receiving photons from a new transmitter, and they realign their Mach–Zehnder interferometers to match the transmitter's interferometer. Then they begin to develop new key material by performing the BBN QKD protocols with this new transmitter.

---

*For some time, we ran traditional sifting during weekdays and Geneva sifting over the weekends to gain realistic experience with both.

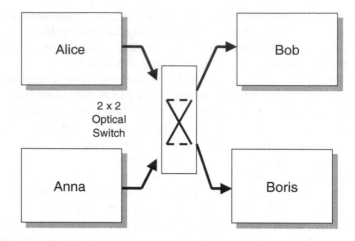

Figure 4.9   Current topology of the DARPA Quantum Network.

The switch chosen for this network is a standard telecommunications facilities switch that operates by moving reflective elements that change the internal light path to produce either a BAR or a CROSS connection. It is operated by applying a TTL-level pulse to either the BAR or the CROSS pin for 20 ms and latches in the activated position. Switching time is 8 ms and optical loss is < 1 dB. Figure 4.10 shows a photograph of the switch mounted on a PC board with the electrical interconnects on the left and optical interconnects on the right.

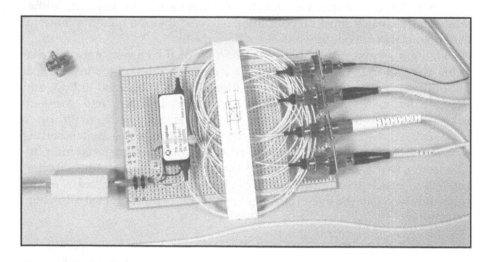

Figure 4.10   The 2 × 2 optical switch mounted on a PC board.

4.8 BBN Key Relay Protocols for Trusted Networks

When two QKD endpoints do not have a direct or photonically switched QKD link between them, but there is a path between them over QKD links through trusted relays, novel BBN-designed networking protocols allow them to agree upon shared QKD bits. They do so by choosing a path through the network, creating a new random number R, and essentially sending R one-time-pad encrypted across each link. We call this process "key relay" and the resultant network a "trusted network" since the chief characteristic of this scheme is that the secrecy of the key depends not just on the endpoints being trustworthy; the intermediate nodes must also be trusted.

The BBN key relay protocols have been continuously operational in the DARPA Quantum Network since October 2003. In fact, they run continuously in our network through Cambridge and allow Alice to build up a reservoir of shared key material with Anna, even though both entities are transmitters, via a trusted relay at Bob or Boris. Similarly Bob and Boris continuously build up shared key material via trusted relays at Alice or Anna.

The four main aspects of the key relay process are illustrated in Figure 4.11. Figure 4.11(A) makes it clear that a key relay network is parallel to an overarching network conveying communication messages and control traffic such as the QKD protocols. Here the Internet is the communications network, each link underneath it is a separate QKD link, and circular nodes are key relay stations. In Figure 4.11(B), one particular source QKD endpoint (S) wishes to agree upon key material with a far-away destination QKD endpoint (D). Since both endpoints, S and D, are connected to a ubiquitous communications

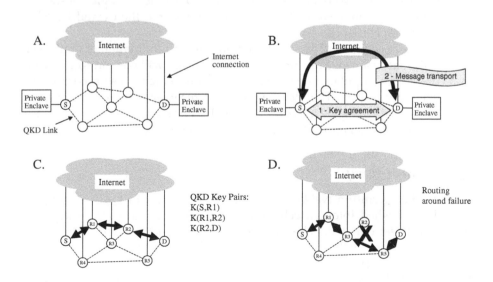

Figure 4.11 Major aspects of the key relay protocols.

network, they can perform QKD protocols in order to derive key material, and once they have agreed upon these keys, they can use the Internet to communicate between themselves securely.

Figure 4.11(C) shows a path used for key relay from S to D, as darkened lines across the key relay network, and the resultant pairs of QKD key material at the right. One QKD-derived key is shared between S and R1; this key is denoted K(S,R1). Likewise K(R1,R2) denotes the pairwise key shared between relay nodes R1 and R2, and so forth. Once all these pairwise keys are in place, S and D can easily derive their own end-to-end shared secret key by key relay. One obvious means is for node S to create a new random number R, protect this number R by K(S,R1), and transmit the result to R1. Node R1 can then decrypt this message to obtain R itself and re-encrypt it by K(R1, R2) to send it onwards to R2, who can in turn repeat the process, and so forth, until it has been relayed all the way to D. At this point, both S and D know the same secret random sequence, R, and can use this shared value as key material.

Finally, Figure 4.11(D) shows that the BBN key relay protocols can automatically discover failures along the key relay path — whether due to cut fiber or eavesdropping — and route the key material around these failures.

4.9   Future Plans

Our near-term plans call for augmenting the DARPA Quantum Network with four new QKD nodes, one pair based on entangled photons in fiber and the other on polarized photons in a free-space channel.

> The entangled link's optical subsystem has been designed and shaken down by Boston University and is now resident at BBN's laboratory. All electronics and software have been built. Once the entangled system is fully operational, we will tie it via key relay into the overall DARPA Quantum Network.

> The free-space link will be based on polarization modulation of faint laser pulses at visible wavelengths. The transmitter will contain four lasers, one for each polarization basis and value, which pulse according to externally supplied random signals; the receiver will perform passive random splitting via a 50/50 coupler. This link will also be woven into the DARPA Quantum Network when operational.

4.10   Summary

The DARPA Quantum Network has married a variety of QKD techniques to well-established Internet technology in order to build a secure key distribution system that can be employed in conjunction with the public Internet or, more likely, with private networks based on the Internet protocol suite. Such private networks are currently in widespread use around the world with customers who desire secure and private communications, e.g., financial institutions, governmental organizations, militaries, and so forth.

The DARPA Quantum Network has been in continuous operation in BBN's laboratory since October 2003, and since June 2004 through dark fiber linking the Harvard University, Boston University, and BBN campuses. It currently consists of four interoperable QKD nodes designed for use through telecommunications fiber and a passive photonic switch that interconnects them; a high-speed free-space system designed and built by NIST; and a full suite of production-quality QKD protocols running on all nodes. Key material derived from these systems is integrated into the Internet security protocols (IPsec) to protect user traffic.

Acknowledgments

We are deeply indebted to Dr. Mike Foster (DARPA IPTO) and Dr. Don Nicholson (Air Force Research Laboratory) who are the sponsor and the agent, respectively, for this research project. We also thank Dr. Carl Williams and his team at NIST for their generous long-term loan of their QKD system. This paper reflects highly collaborative work between the project members. Of these, particular credit is due to Professor Alexander Sergienko, Professor Gregg Jaeger, and Dr. Martin Jaspan (Boston University), Dr. John Myers and Professor Tai Wu (Harvard), and Alex Colvin, John Lowry, William Nelson, David Pearson, Oleksiy Pikalo, John Schlafer, Greg Troxel, and Henry Yeh (BBN). Our interest in QKD networks was sparked by the prior work of the quantum cryptography groups at IBM Almaden and Los Alamos and discussions with them, and by the kind hospitality of Dr. David Murley several years ago.

References

1. C. Elliott, Building the quantum network, *New J. Phys.*, 4, 46, July 2002.
2. R. Hughes, G. Luther, G. Morgan, C. Peterson, and C. Simmons, Quantum cryptography over underground optical fibers, in N. Koblitz, ed., *Advances in Cryptology — CRYPTO '96*, Vol. 1109 of Lecture Notes in Computer Science, Springer-Verlag, 1996, pp. 329–342.
3. G. Brassard, T. Mor, and B. C. Sanders, Quantum cryptography via parametric downconversion, *Quantum Communication, Computing, and Measurement 2: Proceedings of the Fourth International Conference on Quantum Communication, Measurement, and Computing,* Evanston, 1998, P. Kumar, G. Mauro D'Ariano, and O. Hirota, eds., Kluwer Academic/Plenum Publishers, New York, 2000, pp. 381–386.
4. V. Scarani, A. Acin, G. Ribordy, and N. Gisin, Quantum cryptography protocols robust against photon number splitting attacks, ERATO Conference on Quantum Information Science 2003, September 4–6, 2003, Niijima-kaikan, Kyoto, Japan, 2003.

5. A. Kersey, M. Marrone, and M. Davis, Polarization-insensitive fiber optic Michelson interferometer, *Electron. Lett.*, 27, 518, 1991.
6. B. Elliott, O. Pikalo, G. Troxel, and J. Schlafer, Path-length control in an interferometric QKD link, SPIE Aerosense 2002, Proc. SPIE, Vol. 5105.
7. C. Elliott, D. Pearson, and G. Troxel, Quantum cryptography in practice, *Proc. ACM SIGCOMM*, 2003.

chapter 5

Experimental Cryptography Using Continuous Polarization States

S. Lorenz, N. Lütkenhaus, and G. Leuchs
Friedrich Alexander University
of Erlangen-Nürnberg

N. Korolkova
Friedrich Alexander University
of Erlangen-Nürnberg and
University of St. Andrews

Contents

5.1   Introduction..103
5.2   Postselection 107
5.3   Polarization Encoding 109
5.4   Protocol 112
5.5   Sub Shot Noise Polarization Measurement 115
5.6   Alice: Preparation of Polarization States 117
5.7   Signal State Description..118
5.8   Bob: Measurement Systems 119
5.9   Experimental Postselection 121
5.10 Conclusions and Outlook 123
Acknowledgments 123
References 124

5.1   Introduction

Quantum key distribution emerged with an idea and an experiment of Bennett and Brassard in the early 1980s, with the seminal BB84 protocol [1]. At that point it was believed that the only way of rendering the key distribution "quantum" was to use single quanta as information carriers. In the BB84 protocol, the sender Alice prepares the signal states via polarization encoding of single photons, and the receiver Bob decodes them using single photon detection after the appropriately oriented polarizer (for details see [2] and the chapters devoted to the implementation of the BB84 protocol). The protocol relies on the nonorthogonality of the two polarization bases used, i.e., on the impossibility of discriminating deterministically between two nonorthogonal quantum states.

Since the first experiment of Bennett and Brassard [1], the experimental implementation of the BB84 is commonly performed using faint coherent pulses with about 0.1 to 0.4 photons per pulse (for a review see [2]). When using weak coherent pulses, the probability that a single pulse contains more than one photon is not zero. In quantum key distribution (QKD), this was seen for quite some time as a severe drawback as it limits the secure transmission distance [3].

The unconditional security of the BB84 was first proven for the single photon implementation [4] and then also for the faint coherent pulses [5]. Later, a reduced protocol with only two nonorthogonal states, known as the B92 protocol [7], was developed. This paper contains visions which with hindsight carried surprisingly far and include the possibility of continuous variable quantum key distribution with coherent states, which had no immediate impact and was reinvented later (see below). The security of the B92 protocol requires the use of an additional feature compared to the single photon BB84. The signals are encoded using weak nonorthogonal coherent states and a strong reference pulse sent along with each weak signal pulse. The security of the weak signal–strong reference pulse method has still to be rigorously established. However, the idea is that in the intercept–resend strategy, the strong reference pulse prevents the eavesdropper Eve from suppressing unnoticed the intercepted signals, for which she did not manage to gain the desired information. In the absence of the strong reference, this would be a powerful strategy that allows Eve to introduce no errors in the transmission but just reduce the transmission rate which Alice and Bob would then attribute to losses.

The proven possibility of generating a secure key using faint coherent pulses and photon counting has stimulated more thoughts about quantum cryptography with coherent states, leading beyond the early picture of the ultimate importance of single quanta. In 1995, Huttner et al. suggested a so-called $4 + 2$ state protocol [8] combining the strong sides of the BB84 and the B92 protocols. The protocol uses four nonorthogonal coherent states, which are polarization encoded in weak signal pulses and sent along with the intense reference. The detection, however, was still done with a single-photon detector. The preliminary discussion of security aspects in [8] suggests an improved security performance of the $4 + 2$ state protocol as compared to the BB84 (four-state protocol) and the B92 (2-state protocol). An interesting feature of the $4 + 2$ state protocol is that the weak signals and the strong reference are sent as a single pulse in one spatial mode using two orthogonal polarizations. This allows for a convenient practical implementation of the protocol and will later be an important issue for the cryptographic scheme presented below.

Along with coherent states, which are frequently though incorrectly seen as classical, the nonclassical entangled states were also considered for the QKD. The first protocol based on entanglement was suggested by Ekert in 1992 [9]. It uses the test of the Bell inequalities to protect the scheme from an eavesdropper, the invasion of whom would result in entanglement decoherence and can be disclosed by a Bell test of Alice and Bob. The disadvantage is the same as in the previous protocols: stochastic, inefficient sources and inefficient slow detectors. The protocol of Ekert is in its essence equivalent to the BB84 protocol if reformulated in terms of a prepare and measure scheme.

The paper of Ralph [10] marks the starting point of a new era of continuous variable quantum cryptography [10–17]. These continuous variable cryptographic systems all used squeezed or entangled states. The ultimate goal was to achieve high bit rates and thus to solve the problem of low efficiency of discrete variable schemes. As already mentioned, there are no reliable, fast, and deterministically operating single-photon sources available at the moment, and most implementations of single-photon cryptographic systems use weak coherent pulses instead. Owing to the problem of multiphoton components of coherent states in those systems, the effective amplitude has to be very low, which impairs the performance, resulting in a key rate that scales as the square of the single-photon transmission efficiency of the quantum channel. A second drawback is the lack of fast and efficient single-photon detectors, whereas homodyne detectors and bright light photoreceivers work with nearly unit quantum efficiency at high speeds. However, the need for highly nonclassical squeezed or entangled states, the low loss tolerance, and the absence of an unconditional security proof for existing schemes had put the practicability of continuous variable quantum cryptography in question. For a long time it was believed that nonclassical states are the unavoidable prerequisite for protecting the system against eavesdroppers. In addition, the achievable range of key exchange was argued to be restricted to the attenuation length of the channel (50% loss)[18].

The potential advantages in using fast and efficient homodyne detection instead of single-photon counters did not remain unnoticed in the community working with the weak coherent pulse QKD. In 2000 Hirano et al. [19] were the first to replace single-photon detectors with homodyne detection. They used the BB84 protocol with phase encoding and weak coherent pulses using homodyning for detecting quadrature amplitudes. In a way, they progressed on the track pointed out by Bennett [7], that continuous variable cryptography does not necessarily require the use of cumbersome nonclassical states but can get along with quantum coherent states, which are much easier to implement. But they kept the established binary single-photon strategy by extracting binary information from their measured homodyne data. For this purpose they introduced a certain nonzero threshold to differentiate between positive and negative outcomes to provide 0 and 1 bit values. The scheme was still operating on the traditional BB84 with weak coherent pulses. The role of postselection in increasing the ultimate range of continuous variable cryptography has not yet been recognized.

Finally in 2002, Grangier and coworker published a paper [18] proposing QKD using Gaussian modulated coherent states and homodyne detection and showing its security against the beam-splitting attack. This paper reminded the new quantum continuous variable community of the seminal publication by Bennett [7]. The scheme was based on the BB84-type protocol with many nonorthogonal bases represented by slight modulations of different quadrature amplitudes $\hat{x}_\theta = \langle e^{i\theta}\hat{a}^\dagger + e^{-i\theta}\hat{a}\rangle$, which were Gaussian distributed around $\langle \hat{x}_\theta \rangle = 0$. To ensure the sufficient overlap between all signal states, the scheme operated at a low light level, and the modulation depth was kept low enough. The system did not use any postprocessing of the data, and the security of this first scheme was said to be limited to less than 50% loss level, the so-called 3 dB limit. It was argued that this loss limit, implied by the beam-splitting attack, holds for standard minimum uncertainty states such as coherent states as long as no advanced devices such as quantum memories are used.

Shortly after, Silberhorn et al. [20] demonstrated that secure quantum key distribution systems based on continuous variable implementations can operate beyond this apparent 3 dB loss limit. It was shown that, by an appropriate postselection mechanism, reminding one of the approach by Hirano et al. [19], one can enter a region where Eve's knowledge of Alice's key falls behind the information shared between Alice and Bob, even in the presence of substantial losses. The calculations were performed for a particular modification of the protocol of Ref. 18. The security issues related to the postselection in cryptographic schemes using the phase encoded BB84 with weak coherent pulses and homodyne detection were further discussed in Ref. 24.

To overcome the loss limit, another special technique has been proposed by Grangier and coworkers, which uses reverse reconciliation of data [21–23]. The use of reverse reconciliation has demonstrated for the first time the robustness of continuous variable systems against losses of more than 3 dB in an experiment [22]. The original protocol, however, requires strict one-way communication and relies on interferometric stability for the transmission of a local oscillator beam. The restriction to one-way error correction posed a severe limitation, but the same authors showed that certain combinations of one- and two-way communication will also lead to a secure key.

In the following we present the experimental quantum key distribution using coherent polarization states [25]. The implemented system is a continuous variable scheme that combines different features of the traditional discrete variable BB84 [1,8] and continuous variable coherent state cryptography [18–20,22,24]. The distinct difference from the discrete scheme is the use of homodyne detection. The particular properties of our system that make it dissimilar from the related continuous variable schemes [18,19,22,24,26] are polarization encoding and, in contrast to [18,22], the four state protocol based on postselection to ensure security and high loss tolerance.

5.2 Postselection

A coherent state with amplitude α is an eigenstate of the annihilation operator $\hat{a}$ and can be represented as an expansion in the Fock basis with photon number n [27]:

$$|\alpha\rangle = e^{-\frac{|\alpha|^2}{2}} \sum \frac{\alpha^n}{(n!)^{\frac{1}{2}}} |n\rangle. \tag{5.1}$$

Due to the Heisenberg uncertainty principle, a coherent state occupies a certain area in phase space, such that two coherent states $|+\alpha\rangle = e^{i\varphi}|\alpha\rangle$ and $|-\alpha\rangle = e^{i(\varphi+\pi)}|\alpha\rangle$ with $\varphi = 0$ exhibit an overlap

$$f = e^{-2|\alpha|^2}. \tag{5.2}$$

For large $|\alpha|$, this overlap vanishes, and $|+\alpha\rangle$ and $|-\alpha\rangle$ are orthogonal. However, for small $|\alpha|$ a quadrature amplitude measurement $x = \langle e^{i\theta}\hat{a}^\dagger + e^{-i\theta}\hat{a}\rangle$ cannot discriminate deterministically between $|+\alpha\rangle$ and $|-\alpha\rangle$. Measuring of $|+\alpha\rangle$ may yield results that could have been produced by a measurement of a $|-\alpha\rangle$ state and vice versa (Figure 5.1). Let us first assume that the quantum channel connecting Alice with Bob is ideal, so that its transmitivity $\eta = 1$. When a sender (Alice) prepares randomly one of the two nonorthogonal pure states, and a receiver (Bob) guesses which state it was by measuring the amplitude x, his error probability

$$p_e(\eta = 1) = Prob(x < 0|\alpha > 0) + Prob(x > 0|\alpha < 0)$$

$$= \frac{1}{2}\left(1 - \sqrt{1 - f^2}\right) \tag{5.3}$$

is directly linked to the resulting overlap and hence to the amplitude $|\alpha|$. For a channel of transmitivity $\eta < 1$, the states impinging on Bob's detector system are changed to $|+\alpha\rangle = \sqrt{\eta}\,|\alpha\rangle$ and $|-\alpha\rangle = \sqrt{\eta}\,e^{i\pi}|\alpha\rangle$, and the error probability

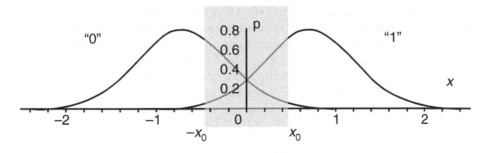

Figure 5.1 Probability distributions of measurement results x for the two nonorthogonal signal states 0 and 1. In the postselection procedure the most inconclusive results $x < |x_0|$ (shaded region) are removed, for which the probability of error is high owing to substantial overlap of the probability distributions $p(0)$ and $p(1)$. The choice of the postselection threshold x_0 is governed by the calculated I_{AB} and I_{AE} as described in the text. (Courtesy of Ch. Silberhorn.)

p_e is modified correspondingly. Figure 5.1 illustrates the dependence of the error probability p_e on x, α, η. The probability distributions for quantum states prepared as 0 and 1 are overlapping, and the degree of the overlap is defined by α, η. The dependence $p_e(x)$ exhibits a certain pattern, which can be exploited in the postselection process taking into account the independence of Bob's and Eve's measurement outcomes x. For Bob's measurement results x falling in the region around the zero point, the error probability of his decision on 0 or 1 on the basis of his measurement is high. On the left and right wings the error probability decreases as the results might be assigned to 0 or 1 with more certainity.

The mutual information between Alice and Bob, I_{AB}, can be determined and depends on Alice's amplitude $|\alpha|$, Bob's amplitude measurement result x, and the transmission η of the channel between Alice and Bob [20,24,28]:

$$I_{AB} = 1 + p_e(x, \alpha, \eta) \, \log_2 p_e(x, \alpha, \eta)$$

$$+ (1 - p_e(x, \alpha, \eta)) \, \log_2(1 - p_e(x, \alpha, \eta)). \tag{5.4}$$

It is not possible yet to give a general statement about the security of the protocol. However, it is possible to show the security of the scheme against beam-splitting attacks, which are the basic kinds of attacks for a potential eavesdropper (Eve) who wants to utilize nonzero quantum channel losses. Eve could split off the part of the signal that is lost in the channel with transmitivity $\eta < 1$. She uses a beam splitter with reflectivity of $1 - \eta$ to obtain that part of Alice's signal that would be lost normally. Eve transmits the rest of the signal over a perfect (lossless) channel, so that her presence is undetectable. She can then make measurements on her part of the signal, e.g., an amplitude measurement like Bob's, and try to infer Bob's measurement results from her own results. For coherent states, Eve's and Bob's probability distributions of measurement outcomes are independent. This means that Eve's and Bob's measurements are completely uncorrelated, so it is possible that Eve obtains inconclusive results while Bob is quite sure about the state Alice prepared, and vice versa. The average information Eve can get depends on Alice's prepared amplitude α and Eve's portion of the signal (in this case $1 - \eta$), giving a mutual information of I_{AE} between Alice and Eve. It can be shown [20] that the mutual information between Alice and Eve depends only on the effective amplitude of the prepared nonorthogonal coherent states and is *independent* of the measurement outcomes x of her and of Bob:

$$I_{AE} = \frac{1}{2}\left(1 + \sqrt{1 - f^2(\alpha, \eta)}\right) \log\left(1 + \sqrt{1 - f^2(\alpha, \eta)}\right)$$

$$+ \frac{1}{2}\left(1 - \sqrt{1 - f^2(\alpha, \eta)}\right) \log\left(1 - \sqrt{1 - f^2(\alpha, \eta)}\right), \tag{5.5}$$

where $f(\alpha, \eta)$ is the overlap of each pair of the four signal states. Alice and Bob can determine the error probability and shared information I_{AB} (Equation (5.4)) for each single event x after the measurement. For a given channel transmission η and state overlap governed by $|\alpha|$, a lower limit for Bob's measurement result x can be given, where $I_{AB} > I_{AE}$. On this basis, Bob decides on the appropriate value of the postselection threshold x_0 to eliminate the most inconclusive results $x < |x_0|$ (see Figure 5.1). As already mentioned, an important point is that for coherent states, Eve's and Bob's probability distributions of measurement outcomes x are independent. Owing to this fact, the shared knowledge between Alice and Bob about the selected events $x > |x_0|$ is larger than that shared between Alice and Eve, allowing a secret key to be distilled (compare Equations (5.4) and (5.5)). The process of sorting out those events with $I_{AB} > I_{AE}$, i.e., the events that are favorable for Alice and Bob after the data have been recorded is called postselection. The detailed description of this procedure can be found in Ref. 20, which includes the derivation of the relevant formulae.

Postselection procedure for a secure key distillation is not limited to coherent state cryptography. It can be extended to enhance the security of continuous variable quantum cryptography [10–17,26] in general. The postselection analysis [20,29] was already applied to one particular QKD scheme [15] based on the entanglement of intense beams. The error probability in (Equation (5.4)) is then determined by the ratio between the signal level and the noise level (signal-to-noise ratio) [10], which depends on the transmitivity of the channel and on the quality of entanglement. It was shown [29] that postselection of the data for such an entanglement-based scheme enables one to sort out the events with $I_{AB} > I_{AE}$ and thus to distill the secret key, even in the presence of high losses.

5.3 Polarization Encoding

Continuous variable cryptography with coherent states and postselection can be implemented with traditional polarization variables of the BB84 protocol. The role of two incompatible nonorthogonal bases is then taken up by the noncommuting quantum polarization variables, Stokes parameters. The hermitian Stokes operators [30,31] are defined as quantum versions of their classical counterparts [32]:

$$\hat{S}_0 = \hat{a}_x^\dagger \hat{a}_x + \hat{a}_y^\dagger \hat{a}_y = \hat{n}_x + \hat{n}_y = \hat{n} \tag{5.6}$$

$$\hat{S}_1 = \hat{a}_x^\dagger \hat{a}_x - \hat{a}_y^\dagger \hat{a}_y = \hat{n}_x - \hat{n}_y \tag{5.7}$$

$$\hat{S}_2 = \hat{a}_x^\dagger \hat{a}_y + \hat{a}_y^\dagger \hat{a}_x \tag{5.8}$$

$$\hat{S}_3 = i \left(\hat{a}_x^\dagger \hat{a}_y - \hat{a}_y^\dagger \hat{a}_x \right) \tag{5.9}$$

where the x and y subscripts label the creation, destruction, and number operators of quantum harmonic oscillators associated with the x and y photon polarization modes, and $\hat{n}$ is the total photon number operator. The creation and destruction operators have the usual commutation relations,

$$\left[\hat{a}_j, \hat{a}_k^\dagger \right] = \delta_{jk} \qquad j, k = x, y. \tag{5.10}$$

The Stokes operator $\hat{S}_0$ commutes with all the others:

$$[\hat{S}_0, \hat{S}_i] = 0 \qquad i = 1, 2, 3 \tag{5.11}$$

but the operators $\hat{S}_1$, $\hat{S}_2$, and $\hat{S}_3$ satisfy the commutation relations of the SU(2) Lie algebra:

$$[\hat{S}_k, \hat{S}_l] = 2i\, \varepsilon_{klm}\, \hat{S}_m. \tag{5.12}$$

Apart from the factor of 2 and the absence of Planck's constant, this is identical to the commutation relation for components of the angular momentum operator. Simultaneous exact measurements of the quantities represented by these Stokes operators are thus impossible in general, and their means and variances are restricted by the uncertainty relations

$$V_2 V_3 \geq |\langle \hat{S}_1 \rangle|^2, \qquad V_3 V_1 \geq |\langle \hat{S}_2 \rangle|^2, \qquad V_1 V_2 \geq |\langle \hat{S}_3 \rangle|^2, \tag{5.13}$$

where V_j is a convenient shorthand notation for the variance $\langle \hat{S}_j^2 \rangle - \langle \hat{S}_j \rangle^2$ of the quantum Stokes parameter $\hat{S}_j$. The angle brackets denote expectation values with respect to the state of interest.

Figure 5.2 witnesses the convenience of using polarization encoding in coherent state cryptography in place of the quadrature encoding traditional for continuous variable schemes. In a measurement of a Stokes parameter S_j, the mode with high photon number $\hat{a}_x$ is used as a phase reference to determine the photon number in the dark mode $\hat{a}_y$ of orthogonal polarization. Note that in conventional homodyne detection, the measurement of conjugate quadratures (e.g., amplitude and phase) of an optical mode normally requires a separate phase reference (local oscillator). The signal and the local oscillator are in two spatially separated modes. Thus the spatial overlap and the phase stability limit the efficiency of such a setup. The use of the quantum polarization of a two-mode coherent state in our cryptographic system provides a clear practical advantage of having its own built-in strong reference

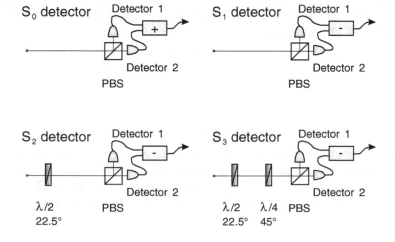

Figure 5.2 Schematical setups to detect all four Stokes parameters. (PBS: polarizing beam splitter.)

field, so that the polarization setup has perfect spatial overlap and a stable relative phase by default without active control.

If in the experiment one of the $\hat{S}_j$ components, e.g., the $\hat{S}_1$ component, is chosen to be large, the quantum Stokes variables effectively acquire the properties of quadrature operators. The arguments used in the section devoted to postselection of coherent states can then be easily extended to the use of coherent polarization states. The choice of "classical" S_1, $|\langle \hat{S}_1 \rangle| \gg 1$, corresponds to an almost completely horizontally polarized light beam with $\hat{S}_2$ and $\hat{S}_3$ as noncommuting observables. The $\hat{S}_2$ and $\hat{S}_3$ measurement bases play the role of two nonorthogonal bases in the BB84 protocol. The S_2 and S_3 components can be measured by applying appropriate phase shifts between $\hat{a}_x$ and $\hat{a}_y$ and using a balanced photodetector (see Figure 5.2) [30]. As seen in Figure 5.2, the switching between these two bases requires the least modification in the measurement setup, which is our motivation to use this particular polarization setting.

Polarization encoding is illustrated in Figure 5.3. A quantum state of a polarized light can be conveniently represented on the Poincare sphere [30]. The quantum state of a p-polarized light is represented with a quantum

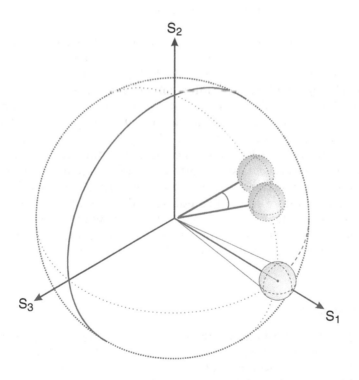

Figure 5.3 Polarization encoding of a coherent p-polarized state on the Poincare sphere. The modulation in S_2 is illustrated for two different modulation amplitudes $(+\delta_1 S_2, +\delta_2 S_2)$.

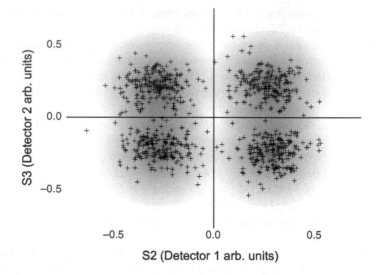

Figure 5.4 Plot of possible Q-function measurement results for Bob. Alice produces four coherent states with either positive or negative S_2 and S_3 polarization. In the experiment, the state overlap is high, so the states cannot be distinguished. In this figure, the overlap is low for better visualization.

uncertainty sphere centered around $(\langle S_1 \rangle, 0, 0)$. For the sake of better visualization, the coherent state with low $\langle S_1 \rangle$ is shown. The modulation in S_2, S_3 results in moving the state around on the Poincare sphere. Modulation in S_2 is depicted in Figure 5.3 by two further coherent states, which differ by some angle θ in the (S_1, S_2) plane corresponding to two modulation amplitudes. Again, for better visualization, states with low overlap between the initial and two modulated states are shown. With growth of $\langle S_1 \rangle$ and increasing overlap, the radius of the Poincare sphere goes to infinity, S_1 can be assumed classical, and three depicted states form in the (S_2, S_3) plane the well-known picture of overlapping nonorthogonal states exactly as one is used for coherent states in quadrature amplitude representation (see also Figure 5.4). Note that, in contrast to the BB84, we do not deal here with two nonorthogonal pairs of orthogonal states, all involved states are nonorthogonal. This difference will explained in greater detail in the next section and is an important feature of our protocol.

5.4   Protocol

The security of the BB84 relies in the first line on the nonorthogonality of the two bases. Does one really need two nonorthogonal bases with two orthogonal signal states each? Would it not suffice to exploit just a few nonorthogonal states [6]? The answer is yes, but certain additional arrangements are required to fix an arising security loophole, as we will describe in the following. The first protocol of this type was the B92 protocol, which was suggested by Bennett

[7] in 1992 and which uses only two nonorthogonal states prepared by Alice. The receiver Bob cannot discriminate between these deterministically but he can apply a generalized measurement using an ancillary system. This kind of measurement, known as positive operator valued measurement (POVM or POM), provides Bob with either a correct answer or with no answer, i.e., with an inconclusive result. Bob will get a bit string of 0's, 1's and ?'s, his 0's and 1's being deterministic results perfectly correlated with those of Alice. Discarding the events corresponding to his ?'s time slots via a public classical channel, Alice and Bob can generate the shared key.

However, in this form the B92 protocol is not secure: eavesdropper Eve could go for an intercept–resend strategy and perform on the intercepted bits the same kind of measurements as Bob. She would then suppress the bits for which she gets inconclusive results and resend only the ones she knows with certainty. This loophole can be closed by sending a strong reference pulse along with a faint signal pulse down the quantum channel. At the receiving side, a weak part of the strong signal is split off and gives in interference with the other weak signal the measurement results. In this scenario, Eve can no longer remain unnoticed; in contrast to a weak signal with 0.1–0.4 photons per pulse, the strong reference part will give a macroscopic signal on Bob's side and cannot be suppressed without being noticed. Whenever the strong signal is present, at least the weak part that is split off from it will enter Bob's remaining measurement device and leads there to a photo detection event with some probability. This means that Eve cannot suppress these events with certainty any more. Hence Eve is forced always to send some signal with the strong pulse and will inevitably introduce errors.

As mentioned in the introduction, in 1995 Huttner et al. extended the BB84 and B92 to the so-called 4 + 2 state protocol [8]. Basically it works with the prepare and measure strategy of the BB84 protocol, but all four states are nonorthogonal. In the QKD run, one of these four nonorthogonal coherent states is polarization encoded using weak coherent pulses. The strong reference needed to support the security of the transmission is sent with the same pulse using two orthogonally polarized modes, one for the signal and the other for the reference. This setting is reminiscent of the Stokes operators of Section 5.3: compare the picture "weak signal and strong reference pulse" with the detection scheme of the Stokes operators (Figure 5.2). By a certain change of the basis, one can represent any polarization state as a dark mode containing information about the quantum polarization state and the mode with a high photon number serving as a phase reference. To illustrate differences and similarities, it is worthwhile to visualize different protocols that use polarization encoding on the Poincare sphere [2,8,25,30]. Notably, the signal states of the 4 + 2 state, together with the respective reference pulses, form on the Poincare sphere of Figure 5.3 exactly the same pattern as described in the previous section if four slight modulations $(+\delta S_2, +\delta S_3)$, $(+\delta S_2, -\delta S_3)$, $(-\delta S_2, +\delta S_3)$, and $(-\delta S_2, -\delta S_3)$ are applied. Thus our protocol described below can be easily reformulated in the established language of weak signal and strong reference pulse, which is a useful tool in security considerations.

The major difference between the $4 + 2$ state protocol and our cryptography scheme is in the receiver: the $4 + 2$ state scheme relies on photon counting, whereas our system makes use of homodyne detection.

The key distribution protocol presented works with a BB84-type prepare and measure strategy. By small modulations of the S_1 polarized cw beam, four coherent states with slightly positive (negative) $\hat{S}_2$ and $\hat{S}_3$ are produced with the overlap chosen small enough to render these states nonorthogonal (cf. Figure 5.3). Alice randomly prepares one of these four states and sends it to Bob. Bob chooses randomly a measurement basis out of S_2 and S_3. Figure 5.4 shows possible measurement outcomes for Bob, when he uses a 50 : 50 beam splitter to measure S_2 on one half and S_3 on the other half of the beam (see also Figure 5.12). By assigning a bit value 1 (0) to a positive (negative) measurement result in both S_2 and S_3 bases, a shared key can be established.

The important constituent of the continuous variable system that forms a distinct difference from conventional single-photon BB84 and $4 + 2$ state protocols is the postselection. It should be intentionally incorporated in the protocol, whereas for single photons it is granted for free by nature: the photons affected by the loss do not arrive at the receiver. They are postselected [20]. In contrast to original BB84, Bob has to choose a postselection threshold x_0 according to the actual key distribution data (Figure 5.1). The optimal threshold is determined by the overlap f and attained information advantage $I_{AB} - I_{AE}$, which means by the amplitude $|\alpha|$ and observed channel transmitivity η (loss level).

Bob discards all his measurement results that did not exceed the postselection threshold. For the remaining measurement results with low p_e, he announces his measurement bases and corresponding time slots through the public channel, so that Alice knows Bob's measurement results with high probability. In the case of vanishing overlap between the states, this procedure is deterministic and hence insecure. An eavesdropper may discriminate between all four states and launch an intercept–resend attack without being noticed. By using states with a considerable overlap, the error probability for Eve increases, while Bob and Alice can postselect favorable events.

It is interesting to establish connections among and within the discrete cryptographic systems and their continuous counterparts and to point out the differences. The tight relation between the four-state BB84 protocol, the two-state B92, and the $4 + 2$ state protocol was already discussed in the beginning of this section. We would like to suggest here a few more analogies. The coherent state continuous variable scheme of Grangier and coworkers [18,22] can be interpreted as the BB84 protocol with many bases and homodyne detection which uses reverse reconciliation to render the secret key generation possible also at high loss level. The scheme of Hirano and Namiki with coworkers [19,24] is similar to the BB84 protocol with two bases but uses homodyne detection and postselection as a reconciliation procedure to support the security. The continuous variable systems [18,19,22,24] encode the signals in quadrature amplitudes. The discrete variable systems use mostly polarization or phase encoding [2]. All these schemes operate at very low light intensities to ensure a sufficient overlap of the signal states. Our scheme can be viewed as the $4 + 2$ protocol but with homodyne detection as the receiver's measurement system and with postselection of data to enhance the security beyond the 3-dB loss limit.

5.5 Sub Shot Noise Polarization Measurement

To detect reliably the polarization with high speed and good efficiency, a homodyne setup as in Figure 5.2 is used. For the measurement of the S_3 parameter, the incoming light polarization is rotated by the appropriate $\frac{\lambda}{2}$ and $\frac{\lambda}{4}$ retarders. A subsequent lens is used to focus the incoming beam on the two photodiodes of the homodyne detector. A high-quality calcite Wollaston polarizer is used to separate the two orthogonal polarization components. It produces a contrast of better than 10^6. Each of two resulting beams is reflected by a mirror onto the photodetector diode. The balanced detection systems use two silicon PIN photodiodes (Hamamatsu S3883), which have a large active area (1 mm diameter), low dark noise, fast response, and a high quantum efficiency (> 0.9 electrons per photon at 810 nm wavelength). The photocurrents are subtracted directly before the net current is converted into a symmetric voltage by a Philips NE5211 transimpedance amplifier, which has a high transimpedance of 28 $k\Omega$ and low dark noise. The low dark noise of diodes and electronics is important to resolve the low quantum noise of the polarization signal. Figure 5.5 shows the electronic noise of the detector compared to the signal noise at an input power of 250 μW. At the sampling frequency of Bob's receiver, 100 kHz, the electronic noise is more than 10 dB below the signal noise.

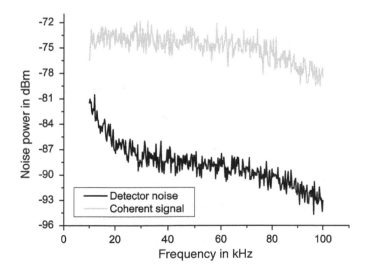

Figure 5.5 Comparision of detector dark noise (lower trace) and detector signal for 500μW input power (upper trace).

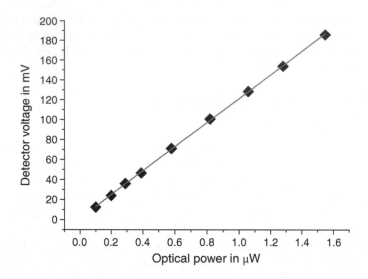

Figure 5.6 DC detector calibration curve. Optical power corresponds to the power on one photodiode, while the other is blocked.

The DC response of the detectors was calibrated by applying a known intensity on one of the two diodes while blocking the other diode. The resulting calibration curve is shown in Figure 5.6. The AC calibration is shown in Figure 5.7. Both diodes were illuminated with the same amount of light, and the AC response was recorded with a rf spectrum analyzer. The detectors

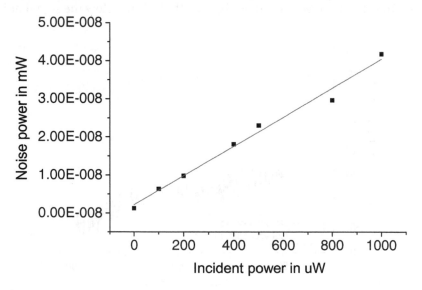

Figure 5.7 AC detector output power vs. light intensity for an unmodulated S_1 beam. This corresponds to a homodyne quadrature measurement of the vacuum field, thus resulting in a linear power dependence of the detected noise signal.

show a linear behavior up to 1 mW incident cw light power. The peak in the frequency response results from the single pole low pass filtering in the detector.

5.6 Alice: Preparation of Polarization States

The experimental apparatus consists of independent setups for Alice and Bob, which are separated by roughly 30 cm. Alice controls the laser source and the state preparation. As a light source a commercial diode laser is used (Figure 5.8). The diode used (TOPTICA DL100) delivers up to 40 mW of power and is wavelength stabilized by an external grating to 810 nm. The laser is decoupled from the rest of the experiment by a tunable optical isolator. A subsequent half-wave plate and polarizing beam splitter combination serves as a variable optical attenuator. The attenuation of roughly 10 dB is used to suppress intensity fluctuations induced by the laser power supply. A mode cleaning of the laser radiation is achieved by coupling the light into a standard telecom fiber. The light that is coupled out of the fiber is then polarized using two polarizing beam splitters, increasing the purity of S_1 polarization. The overall attenuation of the sender setup was adjusted to give an output power of 0.5 mW of S_1 polarized cw light without any modulation (*p*-polarized light). To induce small S_2 and S_3 components in the beam for signal encoding, two modulators were used.

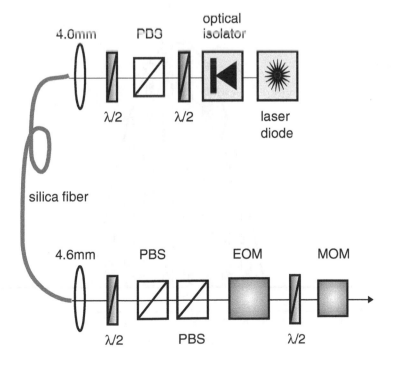

Figure 5.8 Schematic view of Alice's setup.

The S_3 modulation is performed by an electro-optic modulator (EOM). The crystal axis is oriented in a way that small deviations from the quarter-wave voltage result in a circular modulation of the p-polarized laser beam, which corresponds to the modulation in S_3. The S_2 modulation of the beam is achieved by a magneto-optical modulator (MOM), which uses the Faraday effect of a magneto-optically active glass rod (Moltech MOS-04). By applying small longitudinal magnetic fields with a 10 turn copper coil, the active glass rotates the linear polarization, corresponding to an S_2 modulation. By adjusting the EOM voltage and the MOM current, S_2 and S_3 can be modulated continuosly. As the small modulation depth is required to keep the state overlap f large, the voltages applied at EOM and MOM are low. As both modulators are built for free-space optics, their size prohibits high modulation speeds that could be achieved easily with integrated optics. Thus all experiments were done with frequencies in the kHz regime.

5.7   Signal State Description

The modulation pattern of one event is shown in Figure 5.9 for a positive S_2 modulation by Alice. The first and last five samples are zero and are used by Bob to get a reference level. The middle 10 samples give the modulation itself, which can be either positive or negative. In the experiment, both S_2

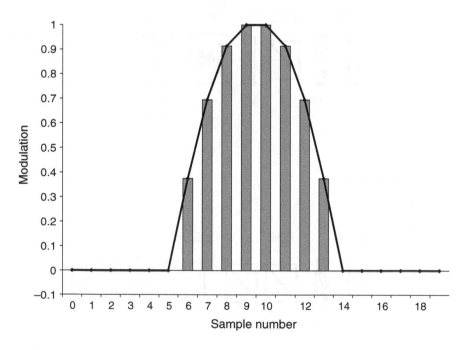

Figure 5.9   Visualization of the positive modulation samples.

and S_3 were modulated randomly with either positive or negative patterns simultaneously. Bob's measurement thus has to discriminate between positive and negative modulation in either of the two polarization bases. To achieve this, Bob records the detector signal for each event and does a baseline restoration using the start and stop parts of the modulation pattern. The central part of his measurements then corresponds to Alice's modulation.

5.8   Bob: Measurement Systems

Bob performs the polarization measurements on the states he receives from Alice (Figures 5.10, 5.11, 5.12). To choose between measurements in S_2 or S_3, Bob can introduce an EOM in his receiver, described in Section 5.6, switching between no retardation for S_2 and quarter-wave retardation for S_3 measurements (see Figure 5.11). Although the EOM has the advantage of introducing little extra loss, it is very slow, owing to its large clear aperture, needed for the free-space transmission line. Thus only low switching speeds (approximately 200 Hz) could be achieved with this type of basis choice.

To circumvent the problem of slow switching speed with the EOM, one can use the loss tolerance advantage of postselection to build a measurement system without active switching. Instead of choosing a basis, both the S_2 and S_3 parameters are measured simultaneously (Figure 5.12). The resulting decrease in accuracy due to the quantum penalty in measurement of two conjugate variables must be encountered by postselection. A polarization-independent beamsplitter is inserted into the beam behind the

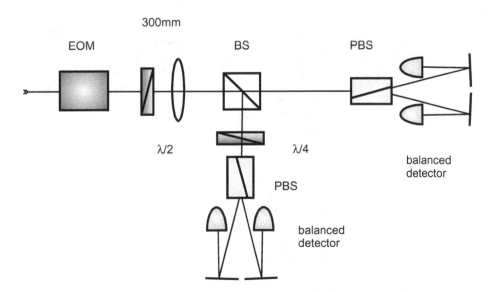

Figure 5.10   Schematic view of Bob's setup, including electro-optical modulator and double detector scheme.

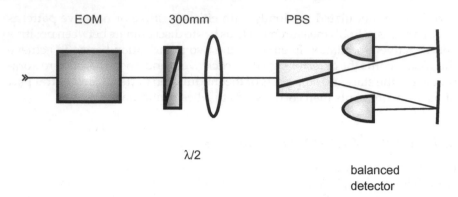

Figure 5.11 Schematic view of Bob's setup, using an electro-optical modulator to change between S_2 and S_3 measurement.

focusing lens (see Figure 5.12). It transmits 52% of the incoming light and reflects 48%. The reflected light is retarded with a quarter-wave plate, to measure its S_3 parameter on a second detector, similar to the one described above. The remaining light is detected by the initial detector to give a measurement of the S_2 parameter. Both detectors can be read out simultaneously. Figure 5.10 shows the combination of both setups that was used in the experiment as a reciever setup enabling us to make different performance tests.

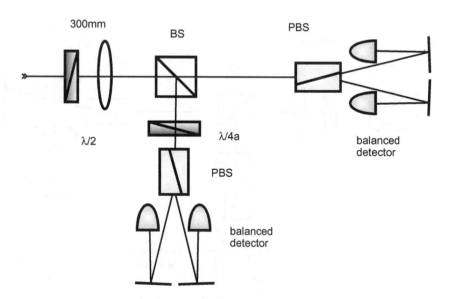

Figure 5.12 Schematic view of Bob's setup, using two detection setups to measure simultaneously S_2 and S_3 with a loss of 50%. If the quarter-wave plate is removed, S_2 can be measured on both halves of the original beams.

5.9 Experimental Postselection

In our demonstration experiment, the setups of Alice (Figure 5.8) and Bob
(Figure 5.10) are separated by roughly 30 cm. To show the independence of
two measurements in the case of a beam splitting attack, the signal is divided at a 50:50 splitter and measured by two Stokes detector setups (see
Figure 5.13). This independence is crucial for the postselection procedure as
described in Section 5.2. When both detectors are set to measure in the same
basis (in this case S_2), an unmodulated S_1 polarized coherent state gives a
Gaussian distribution of the measurement results in each detector. The results are uncorrelated, as can be seen from the circular correlation plot in
Figure 5.13 (left). For an S_2 modulation with low amplitude, the plot (Figure 5.13 right) shows a slight ellipticity, revealing small correlations between
the detected signals. Still it is not possible to deduce the sign of the measurement result of one detector from the outcome of the other detector with
certainty. A potential eavesdropper Eve who uses detector 2 cannot infer the
results of detector 1, even though she has measured 50% of the signal. Note
that the setup of Figure 5.12 was also used with one S_2 and one S_3 detector to
produce Figure 5.4.

In the rest of the experiments, only a single detector with electro-optical
basis switching was used (cf. Figures 5.11 and 5.12). The losses due to nonunity
photodiode efficiency, EOM transmissivity, and optical imperfections in the
detector were treated as if they were transmission channel losses. This conservative point of view implies that Eve could manipulate Bob's receiver and
increase its efficiency while sending more imperfect states at the same time,
gaining additional information. With an additional attenuator in between
Alice and Bob, overall transmission levels of 79% and 36% were set. The
modulation was adjusted to give an average coherent amplitude of 0.6 in the

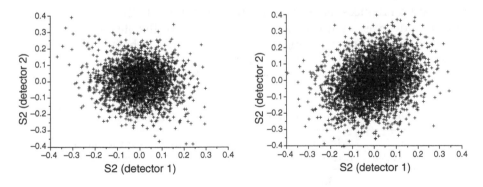

Figure 5.13 Signal divided by a 50:50 beam splitter and measured by two S_2 detectors.
Left: no S_2 modulation of the signal. Right: small S_2 modulation, correlations of the
two detected signals result in elliptical shape of the scatter plot. As detector 1 and
detector 2 have slightly different gains, a small horizontal ellipticity is introduced in
both graphs.

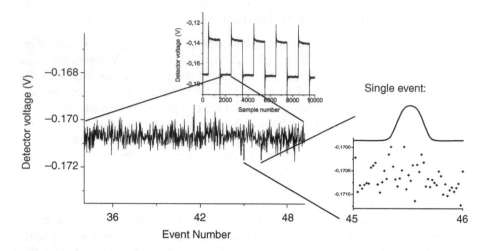

Figure 5.14 Typical oscilloscope trace of the detector signal with constant basis. The insert shows the whole trace, with visible jumps due to basis switching by the EOM.

dark mode d_y, thus giving a considerable overlap according to Equation (5.2). A typical readout of the oscilloscope is shown in Figure 5.14. The main graph shows eleven events with constant measurement basis. In the inset one can see the whole measurement trace, with a basis change visible as a large jump of the signal curve. After the measurement, a bit value was assigned to each of Bob's measurements. In the case of low loss (21%), a postselection threshold of $|\alpha| > 1$ reduced the number of bits from 1001 to 222. The information advantage $I_{AB} - I_{AE}$ was larger than 0.7 bits per event for this threshold. For high losses (64%), the threshold was set to $|\alpha| > 1.5$, giving an information advantage of 0.4 bits per event. 139 out of 956 bits survived the postselection. The error rates for both low and high losses are shown in Table 5.1. Postselection reduces the initial error rate, as expected. As the coherent amplitude α decreases with the losses, the overlap of the states increases (cf. Equation (5.2)). Thus the rising error probability has to be compensated for by error correction, giving also a hint that the channel quality is low. Even though the ratio of secret bits to transmitted bits drops for high losses, it is

Table 5.1 Measurement Results for Low and High Loss

Loss (%)	Inital Errors (%)	Threshold	Residual Errors (%)	Information Advantage
21	15	1	4	0.7
64	22	1.5	10	0.4

Note: Initial errors are before postselection with a threshold as in the third column. After postselection, the fourth column shows the remaining error rate. The information advantage in bits per event is shown in the fifth column.

still possible to select secret information from the measurements. As the post-selection is done by Bob alone, no data has to be transmitted over the public channel during this process. Only after Bob has chosen his events with high information advantage does he reveal their measurement basis to Alice. Thus the effort in authentication and the bandwidth requirements on the public channel decrease substantially.

5.10 Conclusions and Outlook

We have presented an experimental quantum cryptography that uses coherent states, homodyne detection, and postselection to generate a shared key between Alice and Bob. The states are polarization encoded, ensuring fast modulation compared to quadrature encoding and giving a perfect mode overlap in Bob's homodyne detector. There is no need to send a separate local oscillator along with the quantum channel. The system is robust against losses of more than 50% of the states and does not need special reconciliation techniques. Its implementation is very simple, and it could be used in free-space communication with high efficiency and speed. When using a polarization dispersion compensation, the scheme can also be adapted to fiber transmission lines by exchanging the 810 nm coherent laser source with a 1.5 μm laser.

Further improvements could be made with the receiver design. First, the homodyne detection setup of Figure 5.11 can be modified. If one uses the setup described in Section 5.8, used to generate Figure 5.4 (see Figure 5.12), Bob's detector would not require active basis switching, as in some single-photon systems [34]. The theoretical upper bound for detection speed is then the bandwidth of the balanced detector, which can be very high compared to photon counters in single-photon experiments. Second, the photon counting and homodyne systems are not the only possible detection systems. It is worthwhile to look for a system that might be more capable of discriminating nonorthogonal quantum states preserving the advantages of the homodyne technique. For example, one might think about implementation of a modification of the so-called Kennedy receiver [35].

The most important issue is the security of the scheme. Extension of the security analysis to more general kinds of eavesdropping attacks is required. An interesting issue here is the role of squeezing and entanglement in the protection of the system against eavesdroppers. These issues were recently considered by several authors [13,28,29,36].

Acknowledgments

This work was supported by the German Association of Engineers (VDI) and the Federal Ministry of Education and Research (BMBF) under FKZ: 13N8016. The authors would like to thank Ch. Silberhorn for discussions related to postselection procedure and Ulrik Andresen, Jessica Schneider, and Andreas Berger for technical assistance.

References

1. C.H. Bennett, F. Bessette, G. Brassard, L. Salvail, and J. Smolin, Experimental quantum cryptography, *J. Cryptology*, 5, 3, 1992; C.H. Bennett, G. Brassard, and A. Ekert, Quantum cryptography, *Sci. Am.*, 267, 50, 1992.

2. N. Gisin, G. Ribordy, W. Tittel, and H. Zbinden, Quantum cryptography, *Rev. Mod. Phys.*, 74, 145, 2002.

3. G. Brassard, N. Lütkenhaus, T. Mor, and B. C. Sanders, Limitations on practical quantum cryptography, *Phys. Rev. Lett.*, 85, 1330, 2000.

4. D. Mayers, Unconditional security in quantum cryptography, quant-ph/9802025 1998; P. W. Shor and J. Preskill, Simple proof of security of the BB84 quantum key distribution protocol, quant-ph/0003004, 2000.

5. H. Inamori, N. Lütkenhaus, and D. Mayers, Unconditional security of practical quantum key distribution, quant-ph/0107017 2001; N. Lütkenhaus, Security against eavesdropping in quantum cryptography, *Phys. Rev.*, A 54, 97, 200, 1996.

6. C.A. Fuchs, Just two nonorthogonal quantum states, in *Proceedings of the Fourth International Conference on Quantum Communication, Measurement, and Computing, 1998*, G.T. Moore and M. O. Scully, eds., Kluwer Academic, 2000, p. 11.

7. C.H. Bennett, Quantum cryptography using any two nonorthogonal states, *Phys. Rev. Lett.*, 68, 3121, 1992.

8. B. Huttner, N. Imoto, N. Gisin, and T. Mor, Quantum cryptography with coherent states, *Phys. Rev.*, A 51, 1863, 1995.

9. A. Ekert, Quantum cryptography based on Bell's theorem, *Phys. Rev. Lett.*, 67, 661, 1991.

10. T.C. Ralph, Continuous variable quantum cryptography, *Phys. Rev.*, A 61, 010303, 1999; T.C. Ralph, Security of continuous-variable quantum cryptography, *Phys. Rev.*, A 62, 062306, 2000.

11. M. Hillery, Quantum cryptography with squeezed states, *Phys. Rev.*, A 61, 022309, 2000; N.J. Cerf, M. Levy, and G. Van Assche, Quantum distribution of Gaussian keys using squeezed states, *Phys. Rev.*, A 63, 052311, 2001.

12. M.D. Reid, Quantum cryptography with a predetermined key, using continuous-variable Einstein–Podolsky–Rosen correlations, *Phys. Rev.*, A 62, 062308, 2000.

13. D. Gottesman and J. Preskill, Secure quantum key distribution using squeezed states, *Phys. Rev.*, A 63, 022309, 2001.

14. S. Lorenz, Ch. Silberhorn, N. Korolkova, R.S. Windeler, and G. Leuchs, Squeezed light from microstructured fibres: towards free space quantum cryptography, *Appl. Phys.*, B 73, 855, 2001.

15. Ch. Silberhorn, N. Korolkova, and G. Leuchs, Quantum key distribution with bright entangled beams, *Phys. Rev. Lett.*, 88, 167902, 2002.

16. M.G. Raymer and A.C. Funk, Quantum key distribution using nonclassical photon-number correlations in macroscopic light pulses, *Phys. Rev.*, A 63, 022309, 2001.

17. K. Bencheikh, T. Symul, A. Jankovic, and J. A. Levenson, Quantum key distribution with continuous variables, *J. Mod. Opt.*, 48, 1903, 2001.

18. F. Grosshans, P. Grangier, Continuous variable quantum cryptography using coherent states, *Phys. Rev. Lett.*, 88, 057902, 2002.

19. T. Hirano, H. Yamanaka, M. Ashikaga, T. Konishi, and R. Namiki, Quantum cryptography using pulsed homodyne detection, *Phys. Rev.*, A 68, 042331, 2003 and quant-ph/0008037, 2000.

20. C. Silberhorn, T. C. Ralph, N. Lütkenhaus, and G. Leuchs, Continuous variable quantum cryptography: beating the 3 dB loss limit, *Phys. Rev. Lett.*, 89, 167901, 2002.

21. F. Grosshans and P. Grangier, Reversed reconciliation protocols for quantum cryptography with continuous variables, quant-ph/0204127, 2002.

22. F. Grosshans, G. Van Assche, J. Wenger, R. Brouri, N. J. Cerf, and P. Grangier, Quantum key distribution using Gaussian modulated coherent states, *Nature*, 421, 238, 2003.

23. S. Iblisdir, G. Van Assche, N. J. Cerf, Security of quantum key distribution with coherent states and homodyne detection, *Phys. Rev. Lett.*, 93, 170502, 2004.

24. R. Namiki and T. Hirano, Security of quantum cryptography using balanced homodyne detection, *Phys. Rev.*, A 67, 022308, 2003.

25. S. Lorenz, N. Korolkova, and G. Leuchs, Continuous variable quantum key distribution using polarization encoding and postselection, *Appl. Phys.*, B 79, 273, 2004.

26. E. Corndorf, G. Barbosa, Ch. Liang, H.P. Yuen, and P. Kumar, High-speed data encryption over 25 km of fiber by two-mode coherent-state quantum cryptography, *Opt. Lett.*, 28, 2040–2042, 2003; G.A. Barbosa, E. Corndorf, P. Kumar, and H. P. Yuen, Secure communication using mesoscopic coherent states, *Phys. Rev. Lett.*, 90, 227901, 2003.

27. R. Glauber, Coherent and incoherent states of the radiative field, *Phys. Rev.*, 131, 2766, 1963.

28. P. Horak, The role of squeezing in QKD based on homodyne detection and post-selection, quant-ph/0306138, 2003.

29. Ch. Silberhorn, Intensive verschränkte Lichtstrahlen und Quantenkryptographie, Ph.D. thesis, University of Erlangen-Nürnberg, 2003 (in German).

30. N. Korolkova, G. Leuchs, R. Loudon, T. C. Ralph, and Ch. Silberhorn, Polarization squeezing and continuous variable polarization entanglement, *Phys. Rev.*, A, 65, 052306, 2002.

31. B.A. Robson, *The Theory of Polarization Phenomena*, Clarendon Press, Oxford, U.K., 1974.

32. M. Born and E. Wolf, *Principles of Optics*, 7th ed., Cambridge University Press, Cambridge, U.K., 1999.

33. G. Brassard and L. Salvail, in *Advances in Cryptology EUROCRYPT '93*, T. Helleseth, ed., Springer, 1994, Vol. 765 of *Lecture Notes in Computer Science*, pp. 410–423.

34. C. Kurtsiefer, P. Zarda, M. Halder, H. Weinfurter, P. M. Gorman, P. R. Tapster, and J. G. Rarity, Quantum cryptography: A step towards global key distribution, *Nature*, 419, 450, 2002.

35. R.S. Kennedy, A near-optimal receiver for the binary coherent state quantum channel, Quartely Progress Reports QPR, No. 108 of the Research Laboratory of Electronics at the Massachustes Institute of Technology, 1972, p. 219; R. S. Kennedy, On the optimum quantum receiver for the M-ary linearly independent pure state problem, QPR, No. 110, July 1973, p. 142.

36. N. Gisin and N. Brunner, Quantum cryptography with and without entanglement, quant-ph/0312011, 2003.

chapter 6

Quantum Logic Using Linear Optics

J.D. Franson, B.C. Jacobs, and T.B. Pittman
Johns Hopkins University

Contents

6.1   Introduction..127
6.2   Challenges in Quantum Communications..........................128
6.3   Linear Optics Quantum Logic Gates.............................131
6.4   Single-Photon Source and Memory 135
6.5   Quantum Repeaters...139
6.6   Summary...142
Acknowledgments 142
References 142

Abstract

In order for quantum communications systems to become widely used, it will probably be necessary to develop quantum repeaters that can extend the range of quantum key distribution systems and correct for errors in the transmission of quantum information. Quantum logic gates based on linear optical techniques appear to be a promising approach for the development of quantum repeaters, and they may have applications in quantum computing as well. Here we describe the basic principles of logic gates based on linear optics, along with the results from several experimental demonstrations of devices of this kind. A prototype source of single photons and a quantum memory device for photons are also discussed. These devices can be combined with a four-qubit encoding to implement a quantum repeater.

6.1   Introduction

Systems for quantum key distribution have been demonstrated over limited distances, both in optical fibers and in free space, but they have not yet been used for practical applications. In order for quantum communications systems to become widely used, it will probably be necessary to develop quantum repeaters that can extend the range of quantum key distribution systems and correct for errors in the transmission of quantum information. One of the most promising approaches for the development of quantum repeaters is the use of linear optical techniques [1,2] to implement quantum logic gates, combined with optical storage loops to implement a quantum memory device for single photons [3–5]. In this chapter, we describe several prototype quantum logic gates [6–8], a single-photon source [9], and a single-photon memory device [3,4] that we have recently demonstrated. A four-qubit encoding [5] that allows these devices to be combined to implement a quantum repeater will also be described.

The past development of quantum key distribution has been strongly influenced by the need to overcome a variety of practical challenges, and the future development of the field will probably be determined by the challenges that remain. As a result, we begin in Section 6.2 with a brief review of the challenges facing the development of quantum communications systems, both past and future. In Section 6.3, we describe the basic operation of probabilistic quantum logic gates based on linear optics techniques, along with experimental results from several devices of that kind. The development of quantum repeaters will also require a source of single photons and a quantum memory device, and demonstrations of prototype devices of that kind are described in Section 6.4. A proposed implementation [5] of a quantum repeater using a combination of these devices is outlined in Section 6.5, followed by a summary in Section 6.6.

6.2   Challenges in Quantum Communications

Quantum key distribution systems have evolved over the past 15 years in response to a number of technical challenges that limited their performance at the time. As a result, it may be useful to review briefly the past development of quantum key distribution systems and to discuss the remaining challenges that seem likely to determine the future development of the field of quantum communications.

At one time, the only known method for quantum key distribution was based on the use of the polarization states of single photons. In addition to introducing the BB84 and B92 protocols, Bennett et al. also performed the first experimental demonstration of quantum key distribution using photon polarization states in a tabletop experiment [10]. But the use of single-photon polarizations was considered to be a major obstacle to practical applications at the time, since the state of polarization of a photon will change in a time-dependent way as it propagates through an optical fiber. In response to this problem, we developed a feedback loop [11,12] that automatically compensated for the change in polarization of the photons. The system alternated between high-intensity bursts, which determined the necessary corrections to the polarization, and single-photon transmissions, used for the generation of secret key material. The corrections themselves were applied using a set of Pockels cells that also controlled the transmitted polarization state in a BB84 implementation. A system of this kind [13] implemented error correction and privacy amplification in 1994, and it was the first fully automatic and continuously operating quantum key distribution system.

Quantum key distribution systems based on an interferometric approach are now widely used. They have the advantage of being relatively insensitive to changes in the state of polarization in optical fibers. The evolution of interferometer systems of this kind is illustrated in Figure 6.1. The two-photon interferometer shown in Figure 6.1(a) was proposed by one of the authors

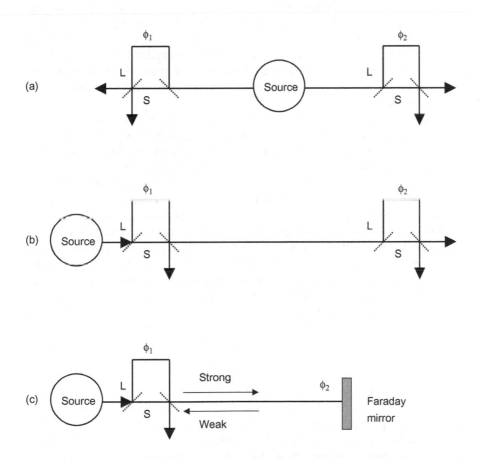

Figure 6.1 Evolution of interferometer-based quantum key distribution systems. (a) Nonlocal interferometer suggested by Franson in which an entangled pair of photons propagate toward two separated interferometers with a long path L and a short path S. (b) Modification by Bennett to utilize a single photon passing through two interferometers in series. (c) Plug-and-play system by Gisin's group that folds the above system in half using a Faraday mirror.

in 1989 [14,15]. Roughly speaking, two entangled photons propagate toward two distant interferometers that both contain a long path L and a short path S. The photons are emitted at the same time in a parametric down-conversion source, and if they arrive at the detectors at the same time, it follows that they both must have traveled the longer path or they both must have traveled the shorter path. Quantum interference between these two probability amplitudes gives rise to nonlocal quantum correlations that violate Bell's inequality.

As early as 1989, John Rarity noted that a two-photon interferometer of this kind could be used as a method of quantum key distribution [16]. Ekert, Rarity, Tapster, and Palma later [17] showed that tests of Bell's inequality could be used to ensure that an eavesdropper cannot determine the polarization states of the photons without being detected, which allows secure communications to be performed. Systems of this kind have now been experimentally demonstrated [18]. One potential advantage of an entangled-photon approach of this kind is that no active devices are required in order to choose a set of random bases for the measurement process. Instead, 50–50 beam splitters can randomly direct each photon toward one of two interferometers with fixed phase shifts.

The interferometric approach of Figure 6.1(a) has the disadvantage of requiring a parametric down-conversion source, which typically has a limited photon generation rate. Charles Bennett realized [19], however, that the need for an entangled source could be eliminated by passing a single photon through two interferometers in series, as illustrated in Figure 6.1(b). Although nonlocal correlations cannot be obtained in such an arrangement, it does allow the use of weak coherent state pulses containing much less than one photon per pulse on the average. The ease in generating weak coherent state pulses combined with the relative lack of sensitivity to polarization changes made this type of interferometer system relatively easy to use. As a result, a number of groups [20–23] demonstrated quantum key distribution systems of this kind, including work by Townsend, Rarity, Tapster, and Hughes.

One of the disadvantages of the interferometric approaches of Figures 6.1(a) and (b) is that the relative phase of the two interferometers must be carefully stabilized. In addition, the polarization of the photons must still be controlled to some extent in order to achieve a stable interference pattern. Gisin and his colleagues [24] avoided both these difficulties by using a very clever technique illustrated in Figure 6.1(c). Here the system is essentially folded in half by placing a mirror at one end of the optical fiber and reflecting the photons back through the same interferometer a second time. By using a Faraday mirror, the state of polarization is changed to the orthogonal state during the second pass through the optical fiber, which eliminates any polarization-changing effects in the optical fiber. Plug-and-play systems of this kind are very stable and are now in widespread use.

The remaining problem in existing quantum key distribution systems is the limited range that can be achieved in optical fibers due to photon loss. As a potential solution to this problem, we performed the first demonstration

[25] of a free-space system over a relatively short distance outdoors in broad daylight in 1996. The accidental detection rate due to the solar background was minimized using a combination of narrow-band filters, short time windows, and a small solid angle over which the signal was accepted. A number of other groups [26,27] have now demonstrated similar systems over larger ranges, and satellite systems of this kind are being considered. These systems will probably have relatively high costs and small bandwidths.

The widespread use of quantum communications systems will require both large bandwidth and operation over large distances. Although earlier limitations due to polarization changes in fibers and the stability of interferometric implementations have now been overcome, it seems likely that quantum repeaters [5,28,29] will be required in order to achieve the necessary bandwidth and operational range. A promising approach for the implementation of a quantum repeater is described in the following sections.

6.3   Linear Optics Quantum Logic Gates

Quantum logic operations are inherently nonlinear, since one qubit must control the state of another qubit. In the case of photonic logic gates, this would seem to require nonlinear optical effects, which are usually significant only for high-intensity beams of light in nonlinear materials. As shown by Knill, Laflamme, and Milburn (KLM), however, probabilistic quantum logic operations can be performed using linear optical elements, additional photons (ancilla), and postselection based on the results of measurements made on the ancilla [1].

The basic idea of linear optical logic gates is illustrated in Figure 6.2. Here two qubits in the form of single photons form the input to the device and two qubits emerge, having undergone the desired logical operation. In addition, a number of ancilla photons also enter the device, where they are combined with the two input qubits using linear optical elements, such as beamsplitters and phase shifters. The quantum states of the ancilla are measured when they leave the device, and there are three possible outcomes: (a) When certain outcomes are obtained, the logic operation is known to have been correctly implemented and the output of the device is accepted without change. (b) When other measurement outcomes are obtained, the output of the device is incorrect, but it can be corrected in a known way using a real-time correction known as feedforward control, which we have recently demonstrated [30]. (c) For the remaining measurement outcomes, the output is known to be incorrect and cannot be corrected using feedforward control. The latter events are rejected and are referred to as failure events. The probability of such a failure can scale as $1/n$ or $1/n^2$, depending on the approach that is used [1,2].

The original approach suggested by KLM was based on the use of nested interferometers [1]. It was subsequently shown [6,31] that similar devices could be implemented using polarization encoding, which had the advantage of simplicity and lack of sensitivity to phase drifts. A controlled NOT (CNOT) quantum logic gate implemented in this way [6] is shown in

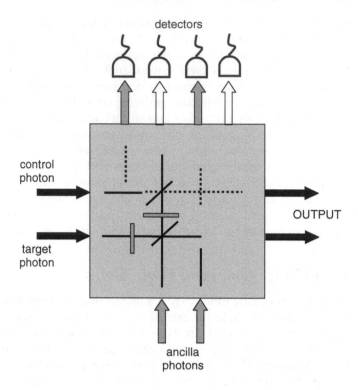

Figure 6.2 Basic idea behind linear optics quantum logic gates. One or more ancillla photons are mixed with two input qubits using linear elements. Postselection based on measurements made on the ancilla will project the correct state of the two output qubits. Feedforward control can be used to accept additional measurement results.

Figure 6.3. Its implementation requires only two polarizing beam splitters, two polarization-sensitive detectors, and a pair of entangled ancilla used as a resource. The correct logical output is obtained whenever each detector registers one and only one photon, which occurs with a probability of 1/4.

The CNOT gate shown in Figure 6.3 can be understood as the combination of several more elementary gates, including the quantum parity check [6,32] shown in Figure 6.4. The intended purpose of this device is to compare the values of the two input qubits without measuring either of them. If the values are the same, then that value is transferred to the output of the device. If the two values are different, then the device indicates that the two bits were different and no output is produced. A quantum parity check of this kind can be implemented using only a single polarizing beam splitter and a single polarization-sensitive detector.

An experimental apparatus [7] used to implement a quantum parity check is outlined in Figure 6.5. Parametric down-conversion was used to generate a pair of photons at the same wavelength. In type-II down-conversion, the two photons have orthogonal polarizations, so that a polarizing beam splitter

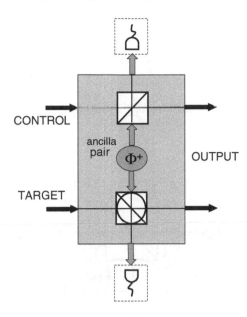

Figure 6.3   Controlled NOT gate using polarization-encoded qubits. The correct logical output is obtained whenever one and only one photon is detected in both detectors, which occurs with a probability of $1/4$.

could be used to separate the photons along two different paths. Waveplates could be used to rotate the plane of polarization of the photons, which created a quantum superposition of logical states, where a horizontally polarized photon represented a value of 0 and a vertically polarized photon represented a value of 1. The parity check itself was implemented with a second polarizing beam splitter, after which the state of polarization could be measured using polarization analyzers and single-photon detectors. The results of the experiment [7] are shown in Figure 6.6 for the case in which the input qubits had definite values of 0 or 1. Here the large data bars correspond to correct results, while incorrect results are seen to be relatively small. Similar performance was also obtained using superposition states as inputs, which demonstrates the quantum-mechanical coherence of the operation.

Another useful quantum logic gate is the quantum encoder [6] shown in Figure 6.7. The intended function of this device is to copy the value of a single input qubit onto two output qubits. Once again, this operation has to be performed without measuring the value of the qubits. Our implementation of a quantum encoder requires a pair of entangled ancilla photons in addition to a polarizing beam splitter. The results from an experimental demonstration [33] of a quantum encoder are shown in Figure 6.8. Once again, the error rate can be seen to be relatively small.

It can be seen that the quantum parity check and encoder form the upper half of our CNOT gate shown in Figure 6.3. The operation of such a device would require four single photons, two of them in an entangled state.

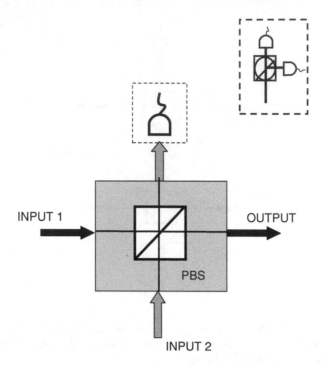

Figure 6.4 Implementation of a parity check operation using a polarizing beam splitter (PBS) and a polarization-sensitive detector. As shown in the insert, the polarization-sensitive detector consists of a second polarizing beam splitter oriented at a 45-degree angle, followed by two ordinary single-photon detectors.

A CNOT operation can also be performed using a three-photon arrangement [8] in which a single ancilla enters the top of the diagram and exits from below, as shown in Figure 6.9. Although this arrangement is easier to implement, the correct results are only obtained when a single photon actually exits in each output port, which can be verified using coincidence measurements (the so-called coincidence basis). The results from the first experimental demonstration [8] of a CNOT gate for photons are shown in Figure 6.10. Here mode mismatch is responsible for most of the incorrect results.

The devices described above succeed with probabilities ranging from $1/4$ to $1/2$. Increasing the probability of success would require the use of larger numbers of ancilla photons [1,2]. In addition to requiring the generation of ancilla photons in entangled states [34], the ancilla must also be detected with high efficiency. In order to avoid these difficulties, we are currently investigating the possibility of a hybrid approach [35] that combines linear optical techniques with a small amount of nonlinearity. It is expected that an approach of this kind will be able to reduce greatly the requirements for large numbers of ancilla and high detection efficiency. In particular, we have shown that the failure rate of devices of this kind can be reduced to zero using the quantum Zeno effect [35].

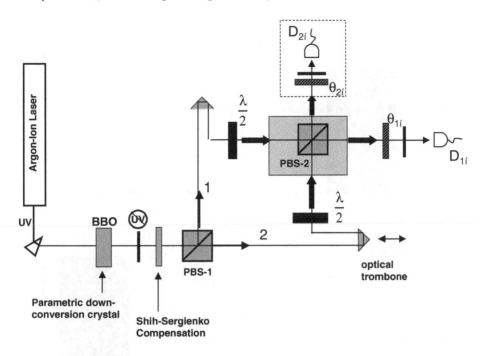

Figure 6.5 Experimental apparatus used to perform a demonstration of a quantum parity check and a destructive CNOT logic gate.

6.4   Single-Photon Source and Memory

The linear optical techniques described above are a promising method for implementing the quantum logic operations that would be required for a quantum repeater. But a source of single photons and a quantum memory would also be required for quantum repeater applications. In this section, we describe prototype experiments in which both of these devices were demonstrated.

In many respects, parametric down-conversion is an ideal way to generate single photons [9]. As illustrated in Figure 6.11, a pulsed laser beam incident on a nonlinear crystal will produce pairs of photons. If one member of a pair is detected, that signals the presence of the other member of the pair. A high-speed optical switch was then used to store the remaining photon in an optical storage loop until it was needed, at which time it could be switched back out of the storage loop. Although a source of this kind cannot produce photons on demand at arbitrary times, it can produce photons at specific times that can be synchronized with the clock time of a quantum computer, which is all that is required for practical applications.

Some experimental results [9] from a single-photon source of this kind are shown in Figure 6.12. It can be seen that the source is capable of producing and storing single photons for later use, but there was a loss of roughly 20%

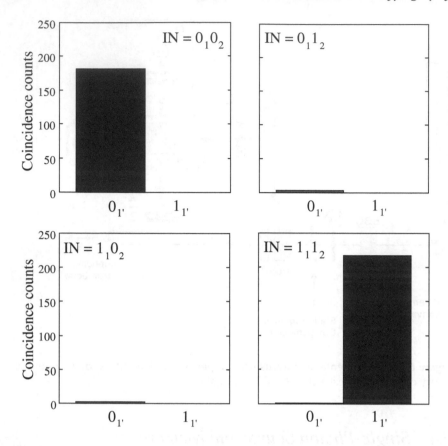

Figure 6.6 Experimental results from a demonstration of a quantum parity check operation. (Reprinted with permission from T.B. Pittman, B.C. Jacobs, and J.D. Franson, Phys. Rev. Lett., 88, 257902, 2002. Copyright 2002 by the American Physical Society.)

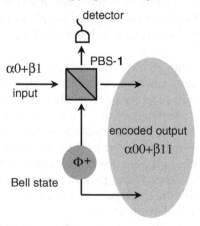

Figure 6.7 Implementation of a quantum encoder using a polarizing beam splitter (PBS) and an entangled pair of ancilla photons.

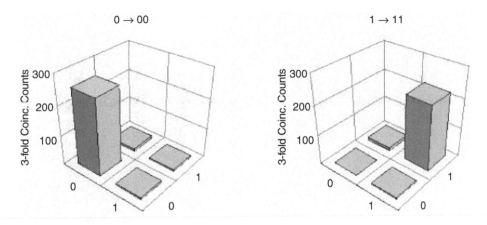

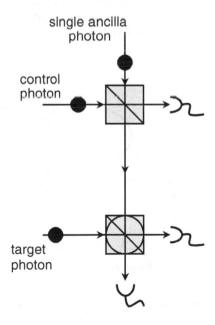

Figure 6.8   Experimental results from a demonstration of a quantum encoder. (Reprinted with permission from T.B. Pittman, B.C. Jacobs, and J.D. Franson, Phys. Rev. A, 69, 042306, 2004. Copyright 2004 by the American Physical Society.)

per cycle time in the original experiment. We are currently working on an improved version of this experiment in which the photons are stored in an optical fiber loop and special-purpose switches are used to reduce the amount of loss.

Figure 6.9   Implementation of a CNOT gate in the coincidence basis using a single ancilla photon. (Reprinted with permission from T.B. Pittman, M.J. Fitch, B.C. Jacobs, and J.D. Franson, Phys. Rev. A, 68, 032316, 2003. Copyright 2003 by the American Physical Society.)

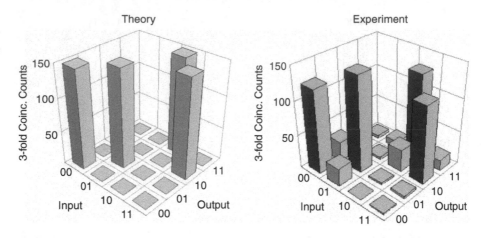

Figure 6.10 First experimental demonstration of a CNOT gate for single photons. (Reprinted with permission from T.B. Pittman, M.J. Fitch, B.C. Jacobs, and J.D. Franson, Phys. Rev. A, 68, 032316, 2003. Copyright 2003 by the American Physical Society.)

It is also possible to construct a quantum memory for photons by switching them into an optical storage loop and then switching them out again when needed [3,4]. In this case the system must maintain the polarization state of the photons in order to preserve the value of the qubit, which is more challenging than the single-photon source described above. This can be accomplished

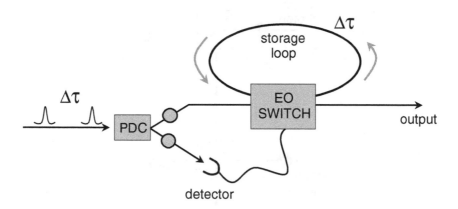

Figure 6.11 Implementation of a single-photon source using pulsed parametric down-conversion (PDC). The detection of one member of a pair of down-converted photons indicates the presence of the second member of the pair, which is then switched into an optical storage loop until it is needed. (Reprinted with permission from T.B. Pittman, B.C. Jacobs, and J.D. Franson, Phys. Rev. A, 66, 042303, 2002. Copyright 2002 by the American Physical Society.)

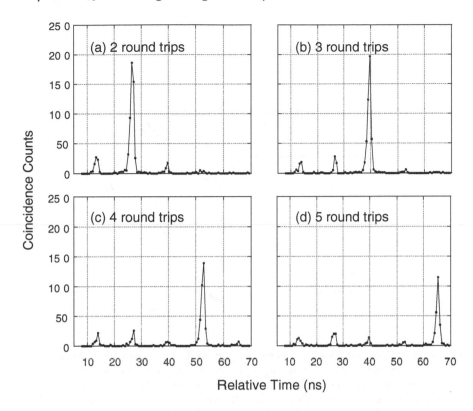

Figure 6.12 Experimental results from a prototype single-photon source. Figures (a) through (d) show the relative probability of switching the photon out after one through five round trips through the optical storage loop. (Reprinted with permission from T.B. Pittman, B.C. Jacobs, and J.D. Franson, Phys. Rev. A, 66, 042303, 2002. Copyright 2002 by the American Physical Society.)

by using a polarizing Sagnac interferometer as the switching mechanism, as illustrated in Figure 6.13. We have also performed a proof-of-principle experiment [4] of this kind where, once again, there were significant losses due to the optical switch.

6.5   *Quantum Repeaters*

In the ideal case, a quantum repeater should be able to correct for all forms of errors that may occur in the transmission of a photon through an optical fiber, including phase and bit-flip errors. But as a practical matter, the dominant error source in fiber based QKD systems is simply the loss of photons due to absorption or scattering. In the quantum key distribution systems that we have implemented, all other sources of error are negligible; there is no measurable decoherence of those photons that pass through the fiber, even when the overall absorption rate is high.

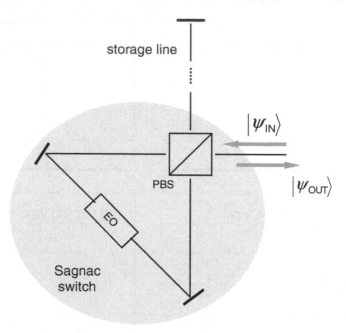

Figure 6.13 Polarizing Sagnac interferometer used as the switching element for a single-photon memory device. A single photon can be stored in the delay line until needed and then switched out again without changing its state of polarization, aside from small technical errors. (Reprinted with permission from T.B. Pittman and J.D. Franson, Phys. Rev. A, 66, 062302, 2002. Copyright 2002 by the American Physical Society.)

As a result, it may be sufficient to consider a quantum repeater system that compensates only for photon loss and simply ignores any other form of error. Such a system can be implemented using a simple four-qubit encoding, as shown by Dowling's group at the Jet Propulsion Laboratory (JPL) [5]. The necessary encoding into four qubits can be done using the circuit shown in Figure 6.14. It can be seen that this encoding can be accomplished using a combination of CNOT logic gates and single-qubit operations, which can be easily implemented in an optical approach.

Once the qubits have been encoded in this way, the effects of photon loss can be corrected [5] using the circuit shown in Figure 6.15. Here a quantum nondemolition measurement is designated by the abbreviation QND; H represents an Hadamard transformation, the sigmas represent the usual Pauli spin matrices and the polygons represent a single-photon source used to replace any photons that have been lost. QND measurements can also be implemented [29,36] using linear optical techniques, so that the entire error correction process can be performed using the kinds of techniques that are described above.

A quantum repeater would then consist of a series of error correction circuits of this kind, separated by a sufficiently short distance L of optical

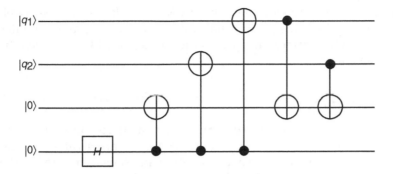

Figure 6.14 Circuit used to encode two logical qubits into four physical qubits, as suggested by Gingrich et al. [5]. (Reprinted with permission from R.M. Gingrich, P. Kok, H. Lee, F. Vatan, and J.P. Dowling, Phys. Rev. Lett., 91, 217901, 2003. Copyright 2003 by the American Physical Society.)

fiber that the probability of absorbing two or more photons in a distance L is negligibly small. Alternatively, the optical fibers could be formed into a set of loops to implement a quantum memory device, as described above, where the error correction circuits would correct for the effects of photon loss and extend the storage time [3,5].

Since the error correction circuit of Figure 6.15 does not correct for other types of errors, it will also be necessary to minimize the failure rate of the CNOT gates by using a sufficiently large number of ancilla photons [2] or by

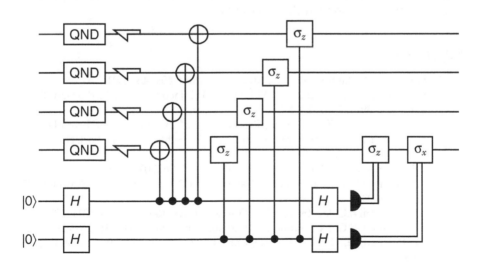

Figure 6.15 A circuit that can be used to correct for photon loss based on the four-qubit encoding of Figure 6.14, as proposed by Gingrich et al. [5]. (Reprinted with permission from R.M. Gingrich, P. Kok, H. Lee, F. Vatan, and J.P. Dowling, Phys. Rev. Lett., 91, 217901, 2003. Copyright 2003 by the American Physical Society.)

using a concatenated code as described by KLM [1]. It may also be possible to reduce the requirements on the number of ancilla and the detector efficiency by using a hybrid approach, such as the Zeno gates [35] that were briefly mentioned above.

6.6   Summary

In summary, we have reviewed some of the challenges faced by quantum communications systems, both past and present. Earlier difficulties associated with changes in the state of polarization and sensitivities to interferometer phase drift have been largely overcome. Although free-space systems will probably be used for special applications, their bandwidth is limited, and quantum repeaters will probably be required in order to achieve the desired bandwidth and operating range. We have demonstrated several kinds of quantum logic gates [6–8], along with a prototype source of single photons [9] and a quantum memory device [4]. As shown by the group at JPL [5], these techniques can be combined with a four-qubit code to correct for the effects of photon loss and to implement a quantum repeater system. Further work will be required in order to reduce the failure rate of linear optics quantum logic gates, possibly including the development of hybrid approaches such as Zeno gates [35].

Acknowledgments

This work was supported by the Army Research Office and Independent Research and Development (IR&D) funds.

References

1. E. Knill, R. Laflamme, and G.J. Milburn, *Nature*, 409, 46, 2001.
2. J.D. Franson, M.M. Donegan, M.J. Fitch, B.C. Jacobs, and T.B. Pittman, Phys. Rev. Lett., 89, 137901, 2002.
3. J.D. Franson and T.B. Pittman, Proceedings of the fundamental problems in quantum theory workshop, D. Lust and W. Schleich, eds., Fortschritte der Physik, 46, 697–705, 1998.
4. T.B. Pittman and J.D. Franson, Phys. Rev. A, 66, 062302, 2002.
5. R.M. Gingrich, P. Kok, H. Lee, F. Vatan, and J.P. Dowling, Phys. Rev. Lett., 91, 217901, 2003.
6. T.B. Pittman, B.C. Jacobs, and J.D. Franson, Phys. Rev. A, 64, 062311, 2001.
7. T.B. Pittman, B.C. Jacobs, and J.D. Franson, Phys. Rev. Lett., 88, 257902, 2002.
8. T.B. Pittman, M.J. Fitch, B.C. Jacobs, and J.D. Franson, Phys. Rev. A, 68, 032316, 2003.
9. T.B. Pittman, B.C. Jacobs, and J D. Franson, Phys. Rev. A, 66, 042303, 2002.
10. C.H. Bennett, F. Bessette, G. Brassard, L. Salvail, and J. Smolin, J. Cryptology, 5, 3, 1992.
11. J.D. Franson and H. Ilves, Appl. Optics, 33, 2949–2954, 1994.

12. J.D. Franson and H. Ilves, J. Mod. Optics, 41, 2391–2396, 1994.
13. J.D. Franson and B.C. Jacobs, Electron. Lett., 31, 232–234, 1995.
14. J.D. Franson, Phys. Rev. Lett., 62, 2205–2208, 1989.
15. J.D. Franson, Phys. Rev. Lett., 67, 290–293, 1991.
16. J. Rarity, private communication, 1989.
17. A.K. Ekert, J.G. Rarity, P.R. Tapster, and G.M. Palma, Phys. Rev. Lett., 69, 1293, 1992.
18. W. Tittel, J. Brendel, H. Zbinden, and N. Gisin, Phys. Rev. Lett., 84, 4737, 2000; G. Ribordy, J. Brendel, J.–D. Gautier, N. Gisin, and H. Zbinden, Phys. Rev. A, 63, 012309, 2001.
19. C.H. Bennett, Phys. Rev. Lett., 68, 3121, 1992.
20. P.D. Townsend, J.G. Rarity, and P.R. Tapster, Electron. Lett., 29, 634, 1993.
21. P.D. Townsend, J.G. Rarity, and P.R. Tapster, Electron. Lett., 29, 1291, 1993.
22. P.D. Townsend, Electron. Lett., 30, 809, 1994.
23. R. Hughes, G. Morgan, and C. Peterson, J. Mod. Optics, 47, 533, 2000.
24. A.T. Muller, T. Herzog, B. Huttner, W. Tittel, H. Zbinden, and N. Gisin, Appl. Phys. Lett., 70, 793, 1997.
25. B. C. Jacobs and J. D. Franson, Optics Lett., 21, 1854–1856, 1996.
26. W.T. Buttler, R.J. Hughes, S.K. Lamoreaux, G.L. Morgan, J.E. Nordholt, and C.G. Peterson, Phys. Rev. Lett., 84, 5652, 2000.
27. J.G. Rarity, P.R. Tapster, and P.M. Gorman, J. Mod. Optics, 48, 1887, 2001.
28. Z. Zhao, T. Yang, Y.-A. Chen, A.-N. Zhang, and J.-W. Pan, Phys. Rev. Lett., 90, 207901, 2003; N. Gisin, G. Ribordy, W. Tittel, H. Zbinden, Rev. Mod. Phys., 74, 145, 2002.
29. B.C. Jacobs, T.B. Pittman, and J.D. Franson, Phys. Rev. A, 66, 052307, 2002.
30. T.B. Pittman, B.C. Jacobs, and J.D. Franson, Phys. Rev. A, 66, 052305, 2002.
31. M. Koashi, T. Yamamoto, and N. Imoto, Phys. Rev. A, 63, 030301, 2001.
32. J.W. Pan, C. Simon, C. Brukner, and A. Zeilinger, Nature, 410, 1067, 2001.
33. T.B. Pittman, B.C. Jacobs, and J.D. Franson, Phys. Rev. A, 69, 042306, 2004.
34. J.D. Franson, M.M. Donegan, and B.C. Jacobs, Phys. Rev. A, 69, 052328, 2004.
35. J.D. Franson, B.C. Jacobs, and T.B. Pittman, Phys. Rev. A, 2004, quant-ph/0408097.
36. P. Kok, H. Lee, and J.P. Dowling, Phys. Rev. A, 66, 063814, 2002.

chapter 7

Practical Quantum Cryptography: Secrecy Capacity and Privacy Amplification

G. Gilbert, M. Hamrick, and F.J. Thayer
Quantum Information Science Group, MITRE*

Contents

7.1   Introduction..146
7.2   Presentation of the Effective Secrecy Capacity......................147
       7.2.1   Secrecy Capacity for Keys of Finite Length 153
7.3   Privacy Amplification: Pointwise Bounds and Average Bounds 154
       7.3.1   Privacy Amplification 155
       7.3.2   Practical Results 156
       7.3.3   Application of Pointwise Bound 158
7.4   Conclusions 160
References 161

Abstract

Quantum cryptography has attracted much attention because of its potential for providing secret communications that cannot be decrypted by any amount of computational effort. Here we provide an analysis of the BB84 quantum cryptographic protocol that simultaneously takes into account and presents the *full* set of analytical expressions for effects due to the presence of pulses containing multiple photons in the attenuated output of the laser, the finite length of individual blocks of key material, losses due to error correction, privacy amplification, and authentication, errors in polarization detection, the efficiency of the detectors, and attenuation processes in the transmission

*Research supported under MITRE Technology program Grant 51MSR 211.

medium. The analysis addresses an extremely important set of eavesdropping attacks on individual photons rather than collective attacks in general. Of particular importance is our derivation of the *necessary and sufficient* amount of privacy amplification compression to ensure secrecy against the loss of key material that occurs when an eavesdropper makes optimized direct (USD), indirect (PNS), and combined individual attacks on pulses containing multiple photons. It is shown that only a fraction of the information in the multiple photon pulses is actually lost to the eavesdropper. We also provide a careful analysis of the use of privacy amplification in quantum cryptography. In order to be practically useful, quantum cryptography must not only provide a guarantee of secrecy but also provide this guarantee with a useful, sufficiently large throughput value. The standard result of generalized privacy amplification yields an upper bound only on the *average value* of the mutual information available to an eavesdropper. Unfortunately this result by itself is inadequate for cryptographic applications. A naive application of the standard result leads one to conclude *incorrectly* that an acceptable upper bound on the mutual information has been achieved. It is the *pointwise value* of the bound on the mutual information, associated with the use of some specific hash function, that corresponds to actual implementations. We provide a fully rigorous mathematical derivation that shows how to obtain a cryptographically acceptable upper bound on the actual, pointwise value of the mutual information. Unlike the bound on the average mutual information, the value of the upper bound on the pointwise mutual information and the number of bits by which the secret key is compressed are specified by two different parameters, and the actual realization of the bound in the pointwise case is necessarily associated with a specific failure probability. The constraints among these parameters, and the effect of their values on the system throughput, have not been previously analyzed. We show that the necessary shortening of the key dictated by the cryptographically correct, pointwise bound, can still produce viable throughput rates that will be useful in practice.

7.1   Introduction

The use of quantum cryptographic protocols to generate key material for use in the encryption of classically transmitted messages has been the subject of intense research activity. The first such protocol, known as BB84 [1], can be realized by encoding the quantum bits representing the raw crytpographic key as polarization states of individual photons. The protocol results in the generation of a shorter string of key material for use by two individuals, conventionally designated Alice and Bob, who wish to communicate using encrypted messages that cannot be deciphered by a third party, conventionally called Eve. The unconditional secrecy of BB84 has been proved under idealized conditions, namely, on the assumption of pure single-photon sources and in the absence of various losses introduced by the equipment that generates and detects the photons or by the quantum channel itself [2]. The conditions under which secrecy can be maintained under more realistic circumstances have been studied extensively [3–6]. Our analysis of the secrecy of a practical implementation of the BB84 protocol simultaneously takes into account and presents the *full* set of analytical expressions for effects due to the presence of pulses containing multiple photons in the attenuated output of the laser, the finite length of individual blocks of key material, losses due to error correction, privacy amplification, and authentication, errors in polarization detection, the efficiency of the detectors, and attenuation processes in the transmission medium [7,13].

We consider particular attacks made on individual photons, as opposed to collective attacks on the full quantum state of the photon pulses. The extension to other protocols, such as B92 [8] is straightforward, but is not discussed here because of limitations of space. We analyze important subtleties that arise in the *practical* implementation of privacy amplification in which the distinction between averaging over hash functions, on the one hand, and making use of a particular hash funtion, on the other, yield different bounds on the mutual information available to an enemy eavesdropper. We pay special attention to the consequences of this distinction on the resulting throughput of secret bits, which is a crucial figure-of-merit in assessing the viability of a practical key distribution system.

7.2 Presentation of the Effective Secrecy Capacity

The protocol begins when Alice selects a random string of m bits from which Bob and she will distill a shorter key of L bits which they both share and about which Eve has exponentially small information. We define the secrecy capacity S as the ratio of the length of the final key to the length of the original string

$$S = \frac{L}{m}. \tag{7.1}$$

This quantity is useful for two reasons. First, it can be used in proving the secrecy of specific practical quantum cryptographic protocols by establishing that

$$S > 0 \tag{7.2}$$

holds for the protocol. Second, it can be used to establish the rate of generation of key material according to

$$\mathcal{R} = \frac{S}{\tau}, \tag{7.3}$$

where τ is the pulse period of the initial sequence of photon transmissions. Several scenarios in which useful key generation rates can be obtained are described in Ref. [7].

The length of the final key is given by

$$L = n - (e_T + q + t + v) - (a + g_{pa}). \tag{7.4}$$

The first term n, is the length of the sifted string. This is the string that remains after Alice has sent her qubits to Bob, and Bob has informed Alice of which qubits were received and in what measurement basis, and Alice has indicated to Bob which basis choices correspond to her own. We consider here the important special case where the number of photons in the pulses sent by Alice follow a Poisson distribution with parameter μ. This is an appropriate description when the source is a pulsed laser that has been attenuated to produce weak coherent pulses. In this case, the length of the sifted string may be expressed as [7]

$$n = \frac{m}{2}\left[\psi_{\geq 1}(\eta\mu\alpha)(1 - r_d) + r_d\right], \tag{7.5}$$

where η is the efficiency of Bob's detector, α is the transmission probability in the quantum channel, and r_d is the probability of obtaining a dark count in Bob's detector during a single pulse period. $\psi_{\geq k}(X)$ is the probability of encountering k or more photons in a pulse selected at random from a stream of Poisson pulses having a mean of X photons per pulse:

$$\psi_{\geq k}(X) \equiv \sum_{l=k}^{\infty} \psi_l(X) = \sum_{l=k}^{\infty} e^{-X}\frac{X^l}{l!}, \tag{7.6}$$

Other types of photon sources may be treated by appropriate modifications of Equations (7.5) and (7.6). A comprehensive treatment of this subject, including an extensive analysis of factors contributing to α, is found in Ref. [7].

The next terms represent information that is either in error or that may be leaked to Eve during the rest of the protocol. This information is removed from the sifted string by the algorithm used for privacy amplification, and so the corresponding number of bits must be subtracted from the length of the sifted string to obtain the size of the final key that results.

The first such term, e_T, represents the errors in the sifted string. This may be expressed in terms of the parameters already defined and the intrinsic channel error probability r_c:

$$e_T = \frac{m}{2}\left[\psi_{\geq 1}(\eta\mu\alpha)r_c(1 - r_d) + \frac{r_d}{2}\right], \tag{7.7}$$

where the intrinsic channel errors are due to relative misalignment of Alice's and Bob's polarization axes and, in the case of fiber optics, the dispersion characteristics of the transmission medium. These errors are removed by an error correction protocol that results in additional q bits of information about the key being transmitted over the classical channel. We express this as

$$q \equiv Q\left(x, \frac{e_T}{n}\right)e_T$$
$$= \frac{xh(e_T/n)}{e_T/n}e_T \tag{7.8}$$

where $h(p)$ is the binary entropy function for a bit whose *a priori* probability of being 1 is p. The factor x is introduced as a measure of the ratio by which a particular error correction protocol exceeds the theoretical minimum amount of leakage given by Shannon entropy [9]:

$$q_{min} = nh(e_T/n) = \frac{h(e_T/n)}{e_T/n} e_T \tag{7.9}$$

The next term, t, is an upper bound for the amount of information Eve can obtain by direct measurement of the polarizations of single-photon pulses. This upper bound can be expressed as

$$t = T e_T \tag{7.10}$$

where T is given by [7,10,11]

$$T(n_1, e_T, e_{T,1}, \epsilon) = \left(\frac{n_1}{e_T} - \frac{e_{T,1}}{e_T} \right) I_{max}^R \left(\frac{e_{T,1}}{n_1} + \xi(n_1, \epsilon) \right)$$

$$+ \xi(n_1, \epsilon) \frac{n_1}{e_T} \left(1 - \frac{e_{T,1}}{n_1} \right)^{1/2}, \tag{7.11}$$

with

$$I_{max}^R(\zeta) \equiv 1 + \log_2 \left[1 - \frac{1}{2} \left(\frac{1 - 3\zeta}{1 - \zeta} \right)^2 \right], \tag{7.12}$$

and ξ is defined by

$$\xi(n_1, \epsilon) \equiv \frac{1}{\sqrt{2n_1}} \text{erf}^{-1}(1 - \epsilon). \tag{7.13}$$

In the above equation ϵ is a security parameter that gives the likelihood for a successful eavesdropping attack against a single-photon pulse in the stream. Finally, we have used

$$n_1 = \frac{m}{2} \left[\psi_1(\eta\mu\alpha)(1 - r_d) + r_d \right] \tag{7.14}$$

and

$$e_{T,1} = \frac{m}{2} \left[r_c \psi_1(\eta\mu\alpha)(1 - r_d) + \frac{r_d}{2} \right], \tag{7.15}$$

which are the contributions to n and e_T from the subset of Alice's pulses for which exactly one photon reaches Bob.

The next term, v, is the information leaked to Eve by making attacks on pulses containing more than one photon. There are a variety of possible attacks, including coherent attacks that operate collectively on all the photons in the pulse. We restrict our attention to disjoint attacks that single out each individual photon. Even with this restriction, there are a number of alternatives. It is not clear that all possible attacks with this restriction have been enumerated in research carried out to date. In this analysis, we consider the situation in which Eve can carry out three important types of attacks. Eve can perform a *direct* attack by making direct measurements of the polarization of some subset of the photons and allowing the rest to continue undisturbed (this is sometimes called an "unambiguous state discrimination" (USD) attack). She can also perform an *indirect* attack by storing some of the photons until she learns Alice's and Bob's basis choices by eavesdropping on their classical channel. She then measures the stored photons in the correct basis to determine unambiguously the value of the bit (this is sometimes called a "photon number splitting" (PNS) attack). Finally, she can make a *combined* attack by using the two strategies in some combination. In Ref. [7] it is shown that the optimum attack is always either a direct or an indirect attack, depending on the value of a parameter y, which depends on channel and detector characteristics and the technological capabilities attributed to Eve [7]. For the case of a fiber optic channel, it is possible in principle for Eve to replace the cable with a lossless medium, so that those pulses whose polarizations she can measure are guaranteed to reach Bob. In this case we take $y = \eta$. For the free-space case, such an attack may not be feasible, but she can achieve a similar effect by using entanglement. In this version of the indirect attack, Eve and an accomplice located near Bob prepare pairs of entangled photons in advance. Eve then entangles one of these pairs with a photon emitted by Alice. Her accomplice can now make measurements on the entangled state, gaining information about the photons at Eve's location without losing photons to the attenuation in the channel. If we allow for such attacks, we still have $y = \eta$. If we do not attribute this level of technology to Eve, it is appropriate to take $y = \eta\alpha$. Note also that Eve can perform direct attacks using classical optical equipment, but that the indirect attacks require the use of a quantum memory.

There are three regions of interest. If $y > 1 - \frac{1}{\sqrt{2}}$ (i.e., $y \gtrsim 0.293$), the indirect attack is stronger, and the maximum information that Eve can obtain is

$$\nu^{\max} = \frac{m}{2} [\![\psi_{\geq 2}(\mu) - (1 - y)^{-1}$$

$$\cdot \{e^{-y\mu} - e^{-\mu}[1 + \mu(1 - y)]\}]\!]. \tag{7.16}$$

If $y < 1 - \frac{1}{3\sqrt{2}}$ (i.e., $y \lesssim 0.206$), the direct attack is stronger, and Eve's information is

$$\nu^{\max} = \frac{m}{2} \left[\psi_2(\mu)y + 1 \right.$$

$$\left. - e^{-\mu} \left(\sqrt{2} \sinh \frac{\mu}{\sqrt{2}} + 2\cosh \frac{\mu}{\sqrt{2}} - 1 \right) \right]. \tag{7.17}$$

Finally, if y lies between these two regions, the relative strength of the attacks depends on the number of photons in the pulse. The information leaked to Eve is

$$\nu^{\max} = \frac{m}{2} \left[\!\!\left[\psi_2(\mu)y + e^{-\mu} \left(\sinh \mu - \sqrt{2} \sinh \frac{\mu}{\sqrt{2}} \right) \right.\right.$$

$$+ \sum_{k=2}^{\infty} \psi_{2k}(\mu) \Big\{ \theta(\sigma_e(k, y) - 1)[1 - (1 - y)^{2k-1}]$$

$$\left.\left. + [1 - \theta(\sigma_e(k, y) - 1)](1 - 2^{1-k}) \Big\} \right]\!\!\right], \tag{7.18}$$

where we have introduced the function

$$\sigma_e(k, y) = \frac{1 - (1 - y)^{2k-1}}{1 - 2^{1-k}}. \tag{7.19}$$

For a photon pulse with $2k$ photons, $\sigma_e(k, y)$ is greater than 1 if the indirect attack is stronger and less than 1 if the direct attack is stronger. For odd numbers of photons, the direct attack is always stronger in this region [7].

The significance of these results for Eve is evident. If the key distribution system is operating in the region of large y, her optimal attack is always the indirect attack. If the system operates in the region of small y, the direct attack is optimal. If the system operates in the middle region, Eve optimizes her attack by measuring nondestructively the number of photons in the incoming pulses and then selecting the attack for each pulse according to the number of photons it contains.

In Figure 7.1 we plot the y-number line, divided into the three optimal attack regions for multiphoton pulses subjected to any of the direct (USD), indirect (PNS) or combined individual attacks. It should be noted that, for many conceivable practical quantum cryptography systems, the values of the relevant parameters are such that one will naturally be located in Region II on the plot, which implies that *the direct (USD) attack is typically going to be stronger than the indirect (PNS) attack.* For instance, a typical system may have photon detectors with efficiencies of about $\eta \simeq 0.5$, and the quantum channel

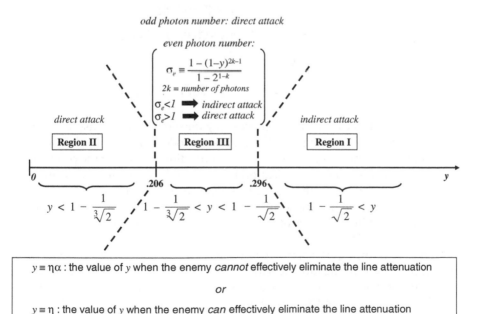

Figure 7.1 Optimal attack regions for multiphoton pulses.

may typically exhibit attenuation values of $\alpha \simeq .01$ (or worse), which yields a value of $y \simeq .005$ (at most), squarely within Region II.

The expressions for ν represent upper bounds on the information that is leaked to Eve by attacks on the individual photons of multiphoton pulses. In Ref. [7] it is shown that Eve can always choose an eavesdropping strategy to achieve this upper bound as long as Bob does not counterattack by monitoring the statistics of multiple detection events that occur at his device. Even with this proviso, the upper bounds are only a fraction of the information contained in the multiphoton pulses. This indicates that the assumption, common in the literature, that Alice and Bob must surrender all this information to Eve is overly restrictive.

The next two terms are grouped together at the end of the expression because their effect on S vanishes in the limit of large m. The first of these, a, is the authentication cost. This is the number of secret bits that are sacrificed as part of the authentication protocol to ensure that the classical transmissions for sifting and error correction occur between Alice and Bob without any "man-in-the-middle" spoofing by Eve. For the authentication protocols described in Ref. [7], the authentication cost is

$$a(n, m) = 4\{g_{auth} + \log_2 \log_2 [2n(1 + \log_2 m)]\}$$

$$\cdot \log_2 [2n(1 + \log_2 m)]$$

$$+ 4[g_{auth} + \log_2 \log_2 (2n)] \log_2 (2n)$$

$$+ 4(g_{EC} + \log_2 \log_2 n) \log_2 n$$

$$+ 4(g_{auth} + \log_2 \log_2 g_{EC}) \log_2 g_{EC}$$

$$+ \tilde{g}_{EC}$$

$$+ 4(g_{auth} + \log_2 \log_2 \tilde{g}_{EC}) \log_2 \tilde{g}_{EC}. \tag{7.20}$$

The first term above is not strictly necessary for security, but is useful in identifying situations in which the authentication process has been compromised. The security parameters g_{auth}, g_{EC}, and $\tilde{g}_{EC}$ are adjusted to limit the probability that some phase of the authentication fails to produce the desired result. For instance, the probability that Eve can successfully replace Alice's transmissions to Bob with her own transmissions is bounded by $2^{-g_{auth}}$. The probability that Alice's and Bob's copies of the key do not match after completion of the protocol is bounded by $2^{-g_{EC}} + 2^{-\tilde{g}_{EC}}$.

The last term, g_{pa}, is a security parameter that characterizes the effectiveness of privacy amplification. It is the number of bits that must be sacrificed to limit the average amount of information, $\langle I \rangle$, about Alice's and Bob's shared key that Eve can obtain to an exponentially small number of bits [12]:

$$\langle I \rangle \leq \frac{2^{-g_{pa}}}{\ln 2}. \tag{7.21}$$

The inequality above furnishes an *average* bound defined with respect to hash functions of a certain type. In specific applications one is necessarily interested in *pointwise* bounds associated with particular hash functions. This is further discussed below.

The fundamental expression for the secrecy capacity may now be written in the limit of small dark count, $r_d \ll 1$:

$$S = \frac{1}{2}\left[\psi_{\geq 1}(\eta\mu\alpha) \cdot (1 - f r_c) + \left(1 - \frac{f}{2}\right) r_d - \tilde{v} \right]$$
$$- \frac{g_{pa} + a}{m}, \tag{7.22}$$

where we have defined

$$f \equiv 1 + Q + T, \tag{7.23}$$

and

$$\tilde{v} \equiv 2 v^{\max}/m, \tag{7.24}$$

so that the rescaled quantity $\tilde{v}$ is independent of m.

Note that the pulse intensity parameter μ can be chosen to maximize the secrecy capacity S and thus also the key generation rate $\mathcal{R}$. A detailed investigation of the optimum pulse intensity under various conditions of practical interest and the resulting secrecy capacities and rates can be found in Ref. [7] and Ref. [13].

7.2.1 Secrecy Capacity for Keys of Finite Length

Most of the terms appearing in Equation (7.4) for the length of the secret key, L, are directly proportional to the length of the block of raw key material, m. After dividing through by m (cf Equation (7.1)), the contributions of these terms to the secrecy capacity S are independent of m. Three of the terms in L are not proportional to m, namely g_{pa}, a, and t. They result in contributions to the effective secrecy capacity that retain explicit dependence on m.

The third contribution, t, requires additional explanation. Its m dependence arises from a precise application of the privacy amplification result, Equation (7.21), derived by Bennett et al. [12]. The bound on Eve's knowledge of the final key is obtained by assuming she has obtained a specific amount of Renyi information prior to privacy amplification. Starting from this point, Slutsky et al. [10] explicitly introduce a security parameter ϵ (see Equation (7.13)) to bound the probability that Eve has obtained more than t bits of Renyi information as a result of her attacks on single-photon pulses.

By contrast, the analysis of Lütkenhaus [3] introduces no parameter analogous to ϵ. Furthermore, the expression for the amount of privacy amplification compression given in Ref. [3] is *linear* in the block size, thus resulting in a contribution to the secrecy capacity that is independent of the block size. While this approach, as developed in Ref. [3], does yield a bound on Eve's information about the key shared by Alice and Bob *after* privacy amplification, explicit results pertaining to the amount of information Eve obtains on the key *prior* to privacy amplification are not presented. Such results have important practical consequences. For example, Eve's likelihood of obtaining more than a given fraction of the raw key from her attacks on single photons increases as the block size of the key material is reduced. One therefore expects that the amount of privacy amplification compression required to ensure secrecy will increase as well. However, since this conclusion is strictly a consequence of the information Eve obtains *prior* to privacy amplification, it cannot *directly* be inferred from the analysis of Ref. [3]. In contrast, the approach of Ref. [10], which we adopt in our analysis, relates the privacy amplification compression directly to the amount of information leaked to Eve *prior* to privacy amplification. This makes it possible to analyze the effect of the block size on the amount of privacy amplification compression, and it concomitantly introduces an explicit security parameter, ϵ, as a bound on Eve's chances of mounting a successful attack on strings of finite length.

7.3   Privacy Amplification: Pointwise Bounds and Average Bounds

Quantum cryptography has been heralded as providing an important advance in secret communications because it provides a guarantee that the amount of mutual information available to an eavesdropper can unconditionally be made arbitrarily small. Any *practical* realization of quantum key distribution that consists only of sifting, error correction, and authentication will allow some information leakage, thus necessitating privacy amplification. Of course, one might contemplate carrying out privacy amplification after executing a classical key distribution protocol. In the absence of any assumed *conditions* on the capability of an eavesdropper, it is not possible to deduce a provable upper bound on the leaked information in the classical case, so that the subsequent implementation of privacy amplification would produce nothing, i.e., the "input" to the privacy amplification algorithm cannot be bounded, and as a result neither can the "output." In the case of quantum key distribution, however, the leaked information associated with the string that is the input to the privacy amplification algorithm can be bounded, and this can be done in the absence of any assumptions about the capability of an eavesdropper. This bound is not good enough for cryptography, however. Nevertheless, this bound on the input allows one to prove a bound on the output of privacy amplification, so that one deduces a final, unconditional upper bound on the mutual information available to an eavesdropper. Moreover, this bound can be made arbitrarily small and hence good enough for cryptography, at the cost of suitably shortening the final string. Except that as usually presented this is not exactly true.

The above understanding is usually presented in connection with the standard result of generalized privacy amplification given by Bennett et al. [12], which applies only to the *average* value of the mutual information. The average is taken with respect to a set of elements, namely, the *universal$_2$* class

of hash functions introduced by Carter and Wegman [14]. The actual implementation of privacy amplification, however, will be executed by software
and hardware that selects a *particular* hash function. The bound on the average value of the mutual information does not apply to this situation: it
does not directly measure the amount of mutual information available to an
eavesdropper in practical quantum cryptography.

In this section we calculate cryptographically acceptable pointwise
bounds on the mutual information that can be achieved while still maintaining sufficiently high throughput rates. In contrast to a direct application
of the privacy amplification result of Ref. [12], we must also consider and
bound a probability of choosing an unsuitable hash function and relate this
to cryptographic properties of the protocol and the throughput rate. The relation between average bounds and pointwise bounds of random variables
follows from elementary probability theory, as was also described in Ref. [15].

7.3.1   Privacy Amplification

In ideal circumstances, the outcome of a k-bit key-exchange protocol is a k-bit
key shared between Alice and Bob that is kept secret from Eve. Perfect secrecy
means that from Eve's perspective the shared key is chosen uniformly from
the space of k-bit keys. In practice, one can only expect that Eve's probability
distribution for the shared key is close to uniform in the sense that its Shannon
entropy is close to its largest possible value k. Moreover, because quantum
key-exchange protocols implemented in practice *inevitably* leak information
to Eve, Eve's distribution of the key is too far from uniform to be usable for
cryptographic purposes. Privacy amplification is the process of obtaining a
nearly uniformly distributed key in a key space of smaller bit size.

We review the standard assumptions of the underlying probability model
of Ref. [12]: Ω is the underlying sample space with probability measure $\mathbf{P}$.
Expectation of a real random variable X with respect to $\mathbf{P}$ is denoted $\mathbf{E}X$. W is
a random variable with key material known jointly to Alice and Bob, and V is
a random variable with Eve's information about W. W takes values in some
finite key space $\mathcal{W}$. The distribution of W is the function $\mathbf{P}_W(w) = \mathbf{P}(W = w)$
for $w \in \mathcal{W}$. Eve's distribution having observed a value v of V is the conditional
probability $\mathbf{P}_{W|V=v}(w) = \mathbf{P}(W = w|V = v)$ on $\mathcal{W}$. In the discussion that follows,
v is fixed, and accordingly we denote Eve's distribution of Alice and Bob's
shared key given v by $\mathbf{P}_{\text{Eve}}$. H and R denote Shannon and Renyi entropies of
random variables defined on $\mathcal{W}$ relative to $\mathbf{P}_{\text{Eve}}$.

Definition 7.1 Suppose $\mathcal{Y}$ *is a key space. If α is a positive real number, a mapping γ :*
$\mathcal{W} \to \mathcal{Y}$ *is an α strong uniformizer for Eve's distribution iff* $H(\gamma) = \Sigma_{y \in \mathcal{Y}} \mathbf{P}_{\text{Eve}}(\gamma^{-1}(y)) \log_2 \mathbf{P}_{\text{Eve}}(\gamma^{-1}(y)) \geq \log_2 |\mathcal{Y}| - \alpha$.

If γ is an α strong uniformizer, then we obtain a bound on the mutual
information between Eve's data V and the image of the hash transformation
Y as

$$I(Y, V) = I(Y) - H(Y|V) = \log_2 |\mathcal{Y}| - H(\gamma) \leq \alpha. \qquad (7.25)$$

Definition 7.2 *Let* Γ *be a random variable with values in* $\mathcal{Y}^W$ *(space of functions* $W \to \mathcal{Y}$*) which is conditionally independent of* W *given* $V = v$*, i.e.,* $\mathbf{P}(\Gamma = \gamma$ *and* $W = w|V = v) = \mathbf{P}(\Gamma = \gamma|V = v)\,\mathbf{P}(W = w|V = v)$*.* Γ *is an* $\alpha > 0$ *average uniformizer for Eve's distribution iff*

$$\mathbf{E}(\mathrm{H}\Gamma) \geq \log_2 |\mathcal{Y}| - \alpha, \tag{7.26}$$

where $\mathrm{H}\Gamma = \mathrm{H}\Gamma(z) = \mathrm{H}(\Gamma(z))$*.*

If Γ is an α average uniformizer, the bound is on the mutual information averaged over the set Γ:

$$I(Y, \Gamma V) = I(Y) - H(Y|\Gamma V) = \log_2 |\mathcal{Y}| - \mathbf{E}(\mathrm{H}\Gamma) \leq \alpha. \tag{7.27}$$

Uniformizers are produced stochastically. Notice that by the conditional stochastic independence assumption, z can be assumed to vary independently of $w \in W$ with the law $\mathbf{P}_{\mathrm{Eve}}$.

Proposition 7.1   *Suppose* Γ *is an* α *average uniformizer. Then for every* $\beta > 0$*,* $\Gamma(\omega)$ *is a* β *strong uniformizer for* ω *outside a set of probability* $\frac{\alpha}{\beta}$*.*

PROOF.   Note that for any $\gamma : W \to \mathcal{Y}$, $\mathrm{H}\gamma$ is at most $\log_2 |\mathcal{Y}|$. Thus $\log_2 |\mathcal{Y}| - \mathrm{H}\Gamma$ is a nonnegative random variable. Applying Chebychev's inequality to $\log_2 |\mathcal{Y}| - \mathrm{H}\Gamma$, it follows that for every $\beta > 0$,

$$\mathbf{P}(\log_2 |\mathcal{Y}| - \beta \geq \mathrm{H}\Gamma) \leq \frac{1}{\beta}\mathbf{E}(\log_2 |\mathcal{Y}| - \mathrm{H}\Gamma)$$

$$= \frac{1}{\beta}(\log_2 |\mathcal{Y}| - \mathbf{E}(\mathrm{H}\Gamma))$$

$$\leq \frac{1}{\beta}\alpha.$$

The random variable Γ is strongly universal$_2$ iff for all $x \neq x' \in X$,

$$\mathbf{P}\{z : \Gamma(z)(x) = \Gamma(z)(x')\} \leq \frac{1}{|\mathcal{Y}|}. \tag{7.28}$$

The following is the main result of Ref. [12]:

Proposition 7.2 (BBCM Privacy Amplification).   *Suppose* Γ *is a universal*$_2$ *family of mappings* $W \to \mathcal{Y}$ *conditionally independent of* W*. Then* Γ *is a* $\frac{2^{\log_2 |\mathcal{Y}| - R(X)}}{\ln 2}$ *average uniformizer for* X*.*

7.3.2   Practical Results

We will refer to the inequality that provides the upper bound on the average value of the mutual information as the *average privacy amplification bound,* or APA, and we will refer to the inequality that provides the upper bound on the actual, or pointwise mutual information as the *pointwise privacy amplifcation bound,* or PPA.

In carrying out privacy amplification, we must shorten the key by the number of bits of information that have potentially been leaked to the eavesdropper [7]. Having taken that into account, we denote by g the additional number of bits by which the key length will be further shortened to assure sufficient secrecy, i.e., the additional bit subtraction amount, and we refer to g as the *privacy amplification subtraction parameter*. With this definition of g, Bennett et al. [12] show as a corollary of Prop. 7.2 that the set of Carter-Wegman hash functions is a $2^{-g}/\ln 2$ average uniformizer. We thus have for the APA bound on $\langle I \rangle$, the average value of the mutual information, the inequality

$$\langle I \rangle \equiv I(Y, \Gamma V) \le \frac{2^{-g}}{\ln 2}. \tag{7.29}$$

In the case of the APA, the quantity g plays a dual role: in addition to representing the number of additional subtraction bits, for the APA case g also directly determines the upper bound on the average of the mutual information.

In the case of PPA we again employ the symbol g to denote the number of subtraction bits, as above for APA, but the upper bound on the pointwise mutual information is now given in terms of a different quantity g', which we refer to as the *pointwise bound parameter*. Also in the case of the PPA we need the parameter g'', which we refer to as the *pointwise probability parameter*, in terms of which we may define the failure probability P_f. This definition is motivated by Prop. 7.1, from which we find that the Carter-Wegman hash functions are $2^{-g}/\ln 2$ strong uniformizers except on a set of probability

$$P_f \le \frac{2^{-g}}{\ln 2} \bigg/ \frac{2^{-g'}}{\ln 2}. \tag{7.30}$$

We therefore define the pointwise probability parameter as

$$g'' \equiv g - g'. \tag{7.31}$$

Thus the quantities g, g' and g'' are not all independent, and are constrained by Equation (7.31). In terms of these parameters we have for the PPA bound on I, the actual value of the mutual information, the inequality

$$I \equiv I(Y, V) \le \frac{2^{-g'}}{\ln 2} = \frac{2^{-(g-g'')}}{\ln 2} \tag{7.32}$$

where the associated failure probability P_f is bounded by

$$P_f \le 2^{-g''}. \tag{7.33}$$

The failure probability is not even a defined quantity in the APA case, but it plays a crucial role in the PPA case. Thus the bound on the pointwise mutual information is directly determined by the value of the parameter g', with respect to which one finds a tradeoff between g, the number of additional compression bits by which the key is shortened, and g'', the negative logarithm of the corresponding failure probability.

7.3.3   Application of Pointwise Bound

Operationally, it will usually be the case in practice that end users of quantum key distribution systems will be first and foremost constrained to ensure that a given upper bound on the pointwise mutual information available to the enemy is realized.

To appreciate the significance of the distinction between the PPA and APA results, we will consider an illustrative example that shows how reliance on the APA bound can lead to complete compromise of cryptographic security. We begin with the APA case. As noted above, in the case of APA, the privacy amplification subtraction parameter, which we will now denote by g_{APA} to emphasize the nature of the bound, directly specifies both the upper bound on $\langle I \rangle$ and the number of bits by which the key needs to be shortened to achieve this bound. Without loss of generality we take the value of the privacy amplification subtraction parameter to be given by $g_{APA} = 30$, which means that, in addition to the compression by the number of bits of information that were estimated to have been leaked, the final length of the key will be further shortened by an additional 30 bits. This results in an upper bound on the average mutual information given by $\langle I \rangle \leq 2^{-30} / \ln 2 \simeq 1.34 \times 10^{-9}$, which we take as the performance requirement for this example. While this might appear to be an acceptable bound, the fact that it applies only to the average of the mutual information of course means that it is not the quantity we require.

We turn to the PPA case, with respect to which we will now refer to the privacy amplification subtraction parameter as g_{PPA}. In order to discuss the PPA bound we must select appropriate values amongst g_{PPA}, g' and g''. In the APA case discussed above, the bound on the (average) mutual information and the number of subtraction bits are both specified by the same parameter g_{APA}. In the PPA case, the number of subtraction bits and the parameter that specifies the bound on the (pointwise) mutual information are not the same. To achieve the same value for the upper bound on I as we discussed for the upper bound on $\langle I \rangle$ above, we must select $g' = 30$ as the value of the pointwise bound parameter. From Equation (7.32), this indeed yields the required inequality $I \leq 2^{-30} / \ln 2 \simeq 1.34 \times 10^{-9}$. However, with respect to this requirement on the value on the mutual information, i.e., the required final amount of cryptographic secrecy, there are a denumerable set (since bits are discrete) of different amounts of compression of the key that are possible to select, each associated with a corresponding failure probability, P_f, in the form of ordered pairs (g_{PPA}, g'') that satisfy the constraint given by $g_{PPA} = g' + g''$ (Equation (7.31)).

Our starting point was the secrecy performance requirement that must be satisfied. On the basis of the APA analysis above, one might conclude that in order to achieve the required secrecy performance constraint it is sufficient to shorten the key by 30 bits. However in the PPA case, satisfying the same performance requirement *and* shortening the key by 30 bits means choosing identical values for the privacy amplification subtraction parameter ($g_{PPA} = 30$) and the pointwise bound parameter ($g' = 30$). However, we note

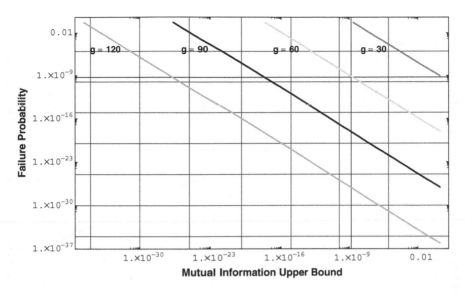

Figure 7.2 Failure probability versus secrecy bound for various privacy amplification compressions.

from Equation (7.31) that in the case of the PPA bound, g_{PPA} and g' become the same only when $g'' = 0$, which corresponds to an upper bound on the failure probability of 100%. (In other words, there is no guarantee that privacy amplification is successful.) This is clearly cryptographically useless!

This example emphasizes the importance of assuring a sufficiently small failure probability in addition to a sufficiently small upper bound on the mutual information. As we see from the above example, the APA result provides no information about the correct number of subtraction bits that are required in order to achieve a specified upper bound on the pointwise mutual information with a suitable failure probability, for which it is essential to use the PPA result instead. In Figure 7.2 we have plotted the failure probability as a function of the upper bound on the mutual information, for a family of choices of g_{PPA} values. Returning to the example discussed above for the APA bound, we see that if we need to achieve an upper bound on I of about 10^{-9}, we may do so with a failure probability of about (coincidentally) 10^{-9}, at the cost of shortening the final key by 60 bits: the secrecy is dictated by the pointwise bound parameter value of $g' = 30$, which is effected by choosing $g_{PPA} = 60$, corresponding to $P_f \simeq 10^{-9}$. Smaller upper bounds can obviously be obtained, with suitable values of the failure probability, at the cost of further shortening of the key.

In Figure 7.3 we plot the throughput of secret Vernam cipher material in bits per second, as a function of bit cell period, for the two bit subtraction amounts $g_{PPA} = 30$ and $g_{PPA} = 60$. The example chosen is a representative scenario for applied quantum cryptography. In calculating the rate we follow

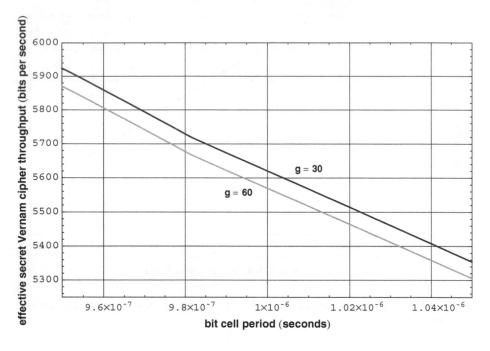

Figure 7.3   Effective throughput of secret Vernam cipher for different *g* values.

the method described in Ref. [7]. We assume the use of an attenuated, pulsed laser, with Alice located on a low Earth orbit satellite at an altitude of 300 kilometers and Bob located at mean sea level, with the various system parameters corresponding to those for Scenario (*i*) in Section 5.3.2 in Ref. [7], except that here the source of the quantum bits operates at a pulse repetition frequency (PRF) of 1 MHz, and we specifically assume that the enemy does not have the capability of making use of prior shared entanglement in conducting eavesdropping attacks. We see that the additional cost incurred in subtracting the amount required to achieve the required mutual information bound and failure probability reduces the throughput rate by an amount that is likely to be acceptable for most purposes. For instance, for a source PRF of 1 MHz, we find that the throughput rate with a value of $g_{PPA} = 30$ is 5614 bits per second. With a subtraction amount of $g_{PPA} = 60$, the throughput rate drops to 5563 bits per second [16].

7.4   Conclusions

We have presented results for the secrecy capacity of a practical quantum key distribution scheme using attenuated laser pulses to carry the quantum information and encoding the raw key material using photon polarizations according to the BB84 protocol. This analysis of the secrecy of a practical implementation of the BB84 protocol simultaneously takes into account and presents the *full* set of analytical expressions for effects due to the presence of pulses containing multiple photons in the attenuated output of the laser, the finite length of individual blocks of key material, losses due to error correction, privacy amplification, and authentication, errors in polarization detection, the efficiency of the detectors, and attenuation processes in the transmission medium for the implementation of BB84 described in Ref. [7]. The transmission medium may be either free space or fiber optic cable. The results apply when eavesdropping is restricted to direct, indirect, and combined attacks on individual photons. The extension of these results to include collective attacks on multiple-photon states in full generality is the subject of continuing research. Of particular importance are the findings that only a portion of the information in the multiphoton pulses need be lost to Eve and the identification of those regions of operation for which Eve's attack is optimized by choosing direct attacks, indirect attacks, or selecting the attack in real time based on the number of photons in the pulse. The assumption, common in the literature, that Alice and Bob must surrender all of this information to Eve is overly conservative. Our analysis presented in Ref. [13] compares quantitatively the results described here for attenuated laser sources with what is achievable using ideal single-photon sources.

The significance and proper implementation of privacy amplification in quantum cryptography are clarified by our analysis. By itself the bound on the average value of the mutual information presented in Ref. [12] does not allow one to determine the values of parameters required to bound the actual, pointwise value of the mutual information. Those parameters must satisfy a constraint, which in turn implies a constraint on the final throughput of secret key material. We have rigorously derived the cryptographically meaningful upper bound on the pointwise mutual information associated with the use of some specific privacy amplification hash function and shown that the corresponding requirements on the shortening of the key still allow viable throughput values.

References

1. C. H. Bennett and G. Brassard, in *Proc. IEEE Int. Conf. on Computers, Systems and Signal Processing*, IEEE Press, New York, 1984.
2. H.-K. Lo, A simple proof of the unconditional security of quantum key distribution, *J.Phys.*, A34, 6957, 2001.
3. N. Lütkenhaus, Estimates for practical quantum cryptography, *Phys. Rev.*, A59, 3301–3319, 1999.
4. G. Brassard, N. Lütkenhaus, T. Mor and B.C. Sanders, Security aspects of practical quantum cryptography, *arXive e-print* quant-ph/9911054, 1999.
5. B. Slutsky, P.-C. Sun, Y. Mazurenko, R. Rao, and Y. Fainman, Effect of channel imperfection on the secrecy capacity of a quantum cryptographic system, *J. Mod. Opt.*, 44, 5, 953–961, 1997.
6. S. Félix, N. Gisin, A. Stefanov, and H. Zbinden, Faint laser quantum key distribution: Eavesdropping exploiting multiphoton pulses, *arXive e-print* quant-ph/0102062, 2001.

7. G. Gilbert and M. Hamrick, Practical quantum cryptography: A comprehensive analysis (part one), *arXive e-print* quant-ph/0009027, 2000.
8. C. H. Bennett, Quantum cryptography using any two nonorthogonal states, *Phys. Rev. Lett.*, 68, 3121, 1992.
9. C.E. Shannon, Communication theory of secrecy systems, *Bell Syst. Tech. J.*, 28, 656, 1949.
10. B. Slutsky, R. Rao, P.-C. Sun, L. Tancevski, and S. Fainman, Defense frontier analysis of quantum cryptographic systems, *App. Opt.*, 37, 14, 2869–2878, 1998.
11. N. Lütkenhaus, Security against individual attacks for realistic quantum key distribution, *Phys. Rev.*, A 61, 052304, 2000.
12. C. H. Bennett, G. Brassard, C. Crépeau, and U. Maurer, Generalized privacy amplification, *IEEE Trans. Inf. Th.*, 41, 1915, 1995.
13. G. Gilbert and M. Hamrick, Secrecy, computational loads and rates in practical quantum cryptography, *Algorithmica*, 34, 314, 2002.
14. J. L. Carter and M. N. Wegman, Universal classes of hash functions, *J. Comp. Syst. Sciences*, 18, 143, 1979.
15. *op. cit.* Lütkenhaus, (ref. [3]). The effect on the viability of throughput rates caused by changing the number of subtraction bits associated with replacing the average bound with the pointwise bound is not analyzed in Ref. [3], and the tradeoffs between the security parameters that define the pointwise bound are not numerically studied. Also, the complete loss of cryptographic security that is caused by naive application of the result given in Ref. [12] is not presented in Ref. [3].
16. The difference between the two throughput values is about 50 bits per second, because an additional 30 bits are subtracted per processing block, and in the example presented there are about 1.6 blocks per second. See Ref. [7] for a discussion of processing block size.

chapter 8

Quantum State Sharing

T. Symul, A.M. Lance, W.P. Bowen, and P.K. Lam
Australian National University

B.C. Sanders
University of Calgary

T.C. Ralph
University of Queensland

Contents

8.1 Introduction 164
8.2 Classical Secret Sharing 165
8.3 Translating Secret Sharing to the Quantum Domain 168
8.4 Implementation of a (2,3) Quantum State Sharing Scheme 170
    8.4.1 The Dealer Protocol 170
    8.4.2 Reconstruction Protocols 172
        8.4.2.1 {1,2} Reconstruction Protocol 172
        8.4.2.2 {2,3} and {1,3} Reconstruction Protocol
                 Using Two OPAs 172
        8.4.2.3 {2,3} and {1,3} Reconstruction Protocol
                 Using a Feedforward Loop 174
    8.4.3 Characterization 176
8.5 Experimental Realization 177
    8.5.1 Experimental Setup 177
    8.5.2 Experimental Results 177
8.6 Applications of Quantum State Sharing 181
    8.6.1 Quantum Information Networks 181
    8.6.2 Quantum Error Correction 183
    8.6.3 Transmission of Entanglement over Faulty Channels 183
    8.6.4 Multipartite Quantum Cryptography 184
Acknowledgments 184
References 184

8.1   Introduction

Cryptography, i.e., the process of scrambling and encoding a plaintext into a cyphertext and then back again, has been in use for centuries. Over time, these techniques have evolved from simple substitution cyphers (where, for example, a letter is replaced with another symbol) to a range of sophisticated mathematical methods due to the advent of computers. Traditionally, cryptography involves only two parties: the *sender* and the *receiver*, commonly know in the literature as *Alice* and *Bob*. In some applications, however, the sender may want to send the secret to more than one receiver so that only through collaboration can a subset of receivers recover it. This situation arises, for example, when the sender cannot trust each receiver individually but can trust a number of receivers collectively. Protocols that facilitate this type of cryptocommunication are known as *secret sharing* protocols. The role of Alice, or the sender, is now replaced by a *dealer* who distributes the secret. The sole receiver, Bob, is now replaced by a number of *players*, none of whom can be completely trusted.

An important class of secret sharing protocols is (k, n) threshold secret sharing [1], in which the dealer encodes and distributes the secret information to n players. Any subset of k players (*the access structure*) must collaborate to retrieve the secret information, while the remaining recipients outside the subset (*the adversary structure*) learn nothing, even when conspiring together. This protocol is widely used to distribute information over classical networks such as the Internet and distributed computer networks.

Quantum state sharing is the quantum equivalent of classical secret sharing, where the classical information is replaced by an unknown quantum state, as illustrated in Figure 8.1. In the ideal case, the access structure can reconstruct the secret quantum state perfectly, even though it appears partially destroyed as a result of malicious and conspiring parties, or catastrophic quantum channel failures.

This chapter is structured in the following way: Section 8.2 gives the reader some background information about classical secret sharing schemes. Section 8.3 discusses how these concepts can be extended to distribute fragile quantum states. It also shows the differences between quantum state sharing

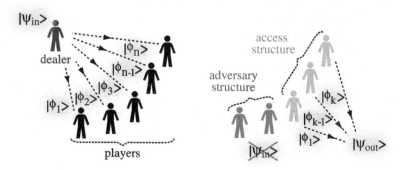

Figure 8.1   Illustration of a quantum state sharing protocol.

and the similarly named, but different, quantum secret sharing protocol proposed by Hillery et al. [2]. In Section 8.4 we discuss how to implement experimentally and characterize a (2,3) threshold quantum state sharing protocol. Section 8.5 shows the first experimental demonstration of a nontrivial threshold quantum state sharing scheme. Finally, Section 8.6 presents an outlook on the applications of this protocol.

8.2   Classical Secret Sharing

The first formal allusion to secret sharing (though not using this nomenclature) was made by Liu [3] through a simple combinatorial problem in the late 1960s. Liu's problem is illustrated in Figure 8.2. Suppose a group of 11

Figure 8.2   The problem of Liu's vault. How many locks on the vault, and how many keys do the scientists (Félix Bloch, Satyendranath Bose, Erwin Schrödinger, Marie Curie, Albert Einstein, David Hilbert, Hendrick Lorentz, Max Planck, Louis de Broglie, Paul Dirac, and Wolfgang Pauli) need to have so that any permutation of six or more of them will be able to open the vault?

scientists are working on a secret project and want to conceal their results in a vault so that none of them can have individual access. The secret document can only be retrieved when a majority of the 11 scientists, that is, any permutation of six or more of them, are present. Liu's vault question is: If a key can only open one lock, what is the minimum number of locks and keys needed to grant this form of access? The answer can be easily calculated to be 462 locks on the vault, and 252 keys per scientist or a total of 2772 keys.

In general where there are n scientists and at least k of them have to collaborate to open the vault, the total number of locks is $C_{k-1}^{n} = n!/((k-1)!(n+1-k)!)$, and each scientist has to carry C_{k-1}^{n-1} keys. It becomes clear from this example that this kind of protocol is impractical when working within the mechanical paradigm. Resource requirements become exponentially large when the number of scientists increases.

Shamir [1] and Blakley [4] independently formalized Liu's problem in 1979. Their goal was to divide some secret information D into n shares D_1, $D_2, \ldots, D_n$ in such a way that

- The knowledge of k or more D_i (where $k \leq n$) makes D easily computable. The ensemble of all the permutations of such shares is referred to as the *access structure*.
- The knowledge of $k - 1$ or fewer D_i gives absolutely no information on D. The ensemble of these shares is known as the *adversary structure*.

Such schemes are called (k, n) *threshold secret sharing*. Shamir's idea is based on polynomial interpolation [1]. The dealer, who wants to share some secret information, chooses a polynomial $P(x)$ of degree $k - 1$, and distributes n pairs of values $(x_1, P(x_1)), (x_2, P(x_2)), \ldots, (x_n, P(x_n))$ to the players. The secret information can be, for instance, the y-axis intercept of the polynomial $P(0)$. It is then clear that any k or more players are able to recover the original polynomial using interpolation of their values (see Figure 8.3(a)). On the other hand, any subset of $k - 1$ or fewer players face an infinite possibility of polynomials that have all their values included. Therefore no information about the secret can be recovered (see Figure 8.3(b)).

Blakley's secret sharing protocol, on the other hand, is based on projective spaces [4]. In his scheme, the dealer distributes a point in a projective space of dimension k to each player. Each of the n players therefore receives a subspace of dimension $k - 1$ (i.e., a hyperplane) comprising the secret data point. k or more players can then intersect their hyperplanes together to retrieve the secret. Similar to Shamir's protocol, $k - 1$ or fewer players will not be able to obtain a unique point by the intersection of their hyperplanes and are thus unable to recover the secret.

Apart from avoiding the explosion of resources required by a mechanical implementation of secret sharing, both Shamir's and Blakley's mathematical methods present several additional advantages. First, the size of information distributed to each player does not exceed the size of the secret. This allows

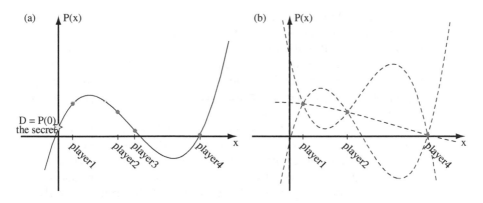

Figure 8.3 Shamir's polynomial $(4, n)$ threshold secret sharing scheme. (a) Four players can collaborate in order to retrieve the original third-order polynomial $P(x)$, and thus the secret information $P(0)$; (b) while three players get infinitely many solutions and are unable to obtain any information.

secret sharing to be efficient. Second, while keeping the secret unchanged, new players can be added or old players deleted without the necessity of modifying the existing access structure. This dynamicity is valuable in many field situations. Third, when the situation arises, it is easy to completely regenerate new shares without changing the secret. All that is needed is to generate a new polynomial or new hyperplanes in projective spaces. Finally, it is possible to improve on the threshold structure by creating a hierarchical access structure where more shares are dealt to a certain subset of players.

Secret sharing, in the context of information technology, is still a very active research area. A lot of new schemes, developments and applications have been proposed. Among them are proposals for proactive secret sharing and verifiable secret sharing. Proactive secret sharing [5] addresses the compromising, loss, or corruption of the shares by periodically renewing the shares without changing the secret. Verifiable secret sharing [6] offers a way for the players to check the authenticity of the shares they receive from the dealer, or the other players from the access structure.

Some applications of secret sharing are

- *The secure access of corporate funding.* In order to secure the money in one of its accounts, the company can use a (k, n) threshold secret sharing protocol to share the account access code among all its authorized employees. In doing so it ensures that at least k stackholders need to collaborate to access the fund and thus lower the risks of having money being stolen, or syphoned off, by any single person or small group.
- *The clipper chip.* This is a National Institute of Standards and Technology proposal to provide the governmental agencies the possibility of

tapping private encrypted communications but still allowing some level of protection against unauthorized tapping [7]. To achieve this, the key used by the private parties to communicate would be shared between two governmental escrow agencies. Both agencies would then need to be contacted independently in order to obtain the cryptographic key to tap the communications, thus providing a basic level of protection to the citizens.

• *Electronic voting.* The privacy of an electronic voting procedure can be ensured by secret sharing [6]. Here is a short explanation on how to proceed. First, the voters need to split their vote into k shares and give them to a group of trusted tallying authorities (who can also verify the identity and uniqueness of the voter). At this point no tallying authority knows what has been voted. These tallying authorities will then count their partial votes independently and send their results to a central authority. Finally, the central authority obtains the final result by concatenating all the partial results together.

8.3   Translating Secret Sharing to the Quantum Domain

As we have seen in the previous section, secret sharing is an important cryptographic protocol designed to distribute secret information to n players, where certain subsets, *the access structure*, can be trusted, and all other subsets, *the adversary structure*, cannot be trusted.

In quantum information processing, the objective in a multiplayer system is not to distribute information but rather to distribute quantum states to the players; hence we employ the term *quantum state sharing* to describe a quantum version of secret sharing. Now the dealer distributes a pure quantum state, or alternatively a mixed state density operator, rather than some classical information to the players. Nevertheless, the properties of classical secret sharing as seen in the previous section are still applicable, except that, as a result of the no-cloning theorem [8], a majority of the players must collaborate to extract the state. This imposes the limitation $n \leq 2k - 1$ on quantum state sharing.

Our employment of the term *quantum state sharing* follows the use of the term *quantum secret sharing* by Cleve et al. [9], which corresponds to the quantum version of secret sharing in cryptography developed by Shamir [1] and Blakley [4]. However, we prefer to use the term *quantum state sharing* to *quantum secret sharing*, as the latter term has been employed for another purpose: protected dissemination of quantum states between completely trustworthy parties in a hostile environment [2,10, 11]. These schemes correspond to our (n, n) threshold quantum state sharing, which is a trivial case of general (k, n) quantum state sharing. Hillery et al. [2] were the first to propose such a scheme using discrete variable GHZ states. They were followed by Karlsson et al. [10], who included entanglement as a resource and also speculated on the possibility of general (k, n) threshold quantum secret sharing. Tittel et al. [11] experimentally realized the (2,2) threshold quantum secret sharing scheme in an elegant experiment involving energy time entangled pseudo-GHZ states. In quantum secret sharing, security against eavesdropper attacks is of paramount importance. Ultimately such techniques offer no security advantage over classical secret sharing used in conjunction with quantum cryptography. Quantum state sharing, on the other hand, is concerned with situations in which the players in the protocol are not themselves completely trustworthy. Such a scheme was proposed by Cleve et al. [9], who gave the full theory for (k, n) threshold quantum state sharing and provided a detailed analysis of the (2,3) threshold quantum state sharing case.

Threshold quantum state sharing, in the spirit of Cleve et al., is interesting for numerous applications. One illustrative example is the distribution of quantum money. One of the first motivations for quantum information theory was Wiesner's suggestion that quantum money could be employed, which was impervious to counterfeiting [12]. Nowadays we have an interest in quantum resources, such as a supply of ebits, or Bell states, but in a sense we can think of this as quantum money since money is a representation of resources. Quantum state sharing allows the distribution of quantum money or quantum resources to multiple players who have to collaborate in predetermined ways (the access structure) in order to use, or spend, this resource. Quantum state sharing allows the quantum money to be locked in a vault until an access structure set with sufficient numbers of keys accesses the vault and removes the quantum money. Other, and probably more crucial, applications of quantum state sharing will be discussed in Section 8.6.

The Cleve et al. result is important as it provides the general protocol for threshold quantum state sharing. The drawback is that it is hard to implement. Even the simplest nontrivial case, namely the (2,3) threshold quantum state sharing scheme, is difficult because it relies on having three qutrits available and the capability of universal transformations on these qutrits. Whereas qutrits and higher order qudits are hard to create and manipulate [13], quantum state sharing with continuous variables is feasible as shown by Tyc and Sanders [14], who developed continuous variable (k, n) threshold quantum state sharing and showed explicitly how to realize the (2,3) threshold quantum state sharing special case. Their scheme utilized Einstein–Podolsky–Rosen (EPR) entanglement [14]. This entanglement is an experimentally accessible quantum resource [15,16] used in quantum information experiments such as continuous variable quantum teleportation [17,18]. This scheme was adapted to a practical scenario by Lance et al. [19], the details of which will be discussed in Sections 8.4 and 8.5. Moreover, Tyc et al. [20] demonstrated that, in the general (k, n) threshold quantum state sharing case, the players never need more than a single EPR-entangled pair, which is an important cost saving for implementation of quantum state sharing protocols.

It has been shown that any (k, n) threshold quantum state sharing schemes with $n \leq 2k - 1$ can be achieved by throwing away shares in a $(k, 2k - 1)$ scheme [9,14]. We can therefore restrict our analysis to the special case of $(k, 2k - 1)$ threshold secret sharing without any loss of generality. We may also restrict our analysis to any set of states that span Hilbert space, since an arbitrary state can be constructed from a linear combination of such states. Here we consider coherent states, since they span Hilbert space and can be readily obtained and manipulated in an experimental setting. A thorough analysis of $(k, 2k - 1)$ threshold quantum state sharing of coherent states can be found in the papers of Tyc et al. [14,20]. To provide the clearest possible analysis, and due to its relevance to the experimental work presented later in this chapter, we restrict ourselves henceforth to the simplest nontrivial case, (2, 3) threshold quantum state sharing.

8.4   Implementation of a (2,3) Quantum State Sharing Scheme

In the continuous variable regime, it is convenient to represent the quantum states using the Heisenberg picture of quantum mechanics. The analysis presented in this chapter will be undertaken in this picture. In particular, we consider states at the frequency sidebands of an electromagnetic field. A quantum state is associated with the field annihilation operator $\hat{a} = (\hat{X}^+ + i\hat{X}^-)/2$, where $\hat{X}^\pm$ are the amplitude (+) and phase (−) quadratures. These quadratures are expanded into steady state $\langle \hat{X}^\pm \rangle$ and fluctuating $\delta \hat{X}^\pm$ components respectively, with $\hat{X}^\pm = \langle \hat{X}^\pm \rangle + \delta \hat{X}^\pm$. The variance of the quadrature operator is given by $V^\pm = \langle (\delta \hat{X}^\pm)^2 \rangle$.

8.4.1   The Dealer Protocol

In the dealer protocol for the (2,3) quantum state sharing scheme proposed by Tyc and Sanders, an entangled state is utilized to encode and distribute the secret quantum state to the players. One way in which this entangled state may be generated is by interfering two amplitude quadrature squeezed beams on a 1:1 beam splitter (x:y beamsplitter with reflectivity $x/(x + y)$ and transmittivity $y/(x + y)$ with a relative phase shift of $\pi/2$, as shown in Figure 8.4. Amplitude squeezed beams are so called because they exhibit an amplitude quadrature noise variance below the quantum noise limit, while correspondingly the phase quadrature has fluctuations over the quantum noise limit. The two output beams resulting from the interference of the amplitude squeezed beams are entangled and can be expressed as

$$\hat{a}_{\text{EPR1}} = (\hat{a}_{\text{sqz1}} + i\hat{a}_{\text{sqz2}})/\sqrt{2} \tag{8.1}$$

$$\hat{a}_{\text{EPR2}} = (i\hat{a}_{\text{sqz1}} + \hat{a}_{\text{sqz2}})/\sqrt{2} \tag{8.2}$$

where $\hat{a}_{\text{sqz1}}$ and $\hat{a}_{\text{sqz2}}$ are the annihilation operators of the amplitude quadrature squeezed beams. The signature of this form of entanglement is that

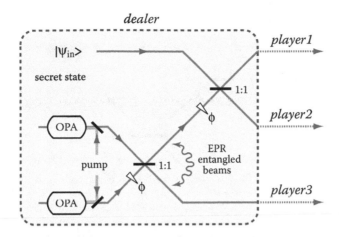

Figure 8.4 Schematc of the dealer protocol for the (2,3) quantum state sharing scheme. OPA: optical parametric amplifier; ψ_{in}: secret quantum state; 1:1: 50% reflectivity beamsplitter; ϕ phase delay.

correlations of both the amplitude and phase quadratures between the beams. We therefore call the state *quadrature entangled*. Such entanglement is directly analogous to the particle position-momentum entanglement described by Einstein, Podolsky and Rosen [21], and as such it can be used to demonstrate the EPR paradox [16].

The dealer generates the players' shares by interfering the secret state $\hat{a}_{in}$ with one of the entangled beams $\hat{a}_{EPR1}$ on a 1:1 beam splitter. The two resulting output fields and the second entangled beam $\hat{a}_{EPR2}$ form the three shares that are distributed to the players, as shown in Figure 8.4. The entangled state ensures that the secret is protected from each player individually. The dealer can further enhance the security of the scheme by displacing the coherent amplitudes of the shares with correlated Gaussian white noise [19]. This additional security can be arbitrarily enhanced by the dealer by increasing the variance of the added Gaussian noise. By choosing the Gaussian noise to have the same correlations as the quadrature entanglement, the shares can then be expressed as

$$\hat{a}_1 = (\hat{a}_{in} + \hat{a}_{EPR1} + \delta N)/\sqrt{2} \tag{8.3}$$

$$\hat{a}_2 = (\hat{a}_{in} - \hat{a}_{EPR1} - \delta N)/\sqrt{2} \tag{8.4}$$

$$\hat{a}_3 = \hat{a}_{EPR2} + \delta N^* \tag{8.5}$$

where $\delta N = (\delta N^+ + i\delta N^-)/2$ represents the Gaussian noise with mean $\langle \delta N^\pm \rangle = 0$ and variance $\langle (\delta N^\pm)^2 \rangle$, and $*$ denotes the complex conjugate. The reconstruction protocol used for the (2,3) quantum state sharing scheme is dependent on which players constitute the access structure. The next section examines the reconstruction protocols used by the different access structure sets.

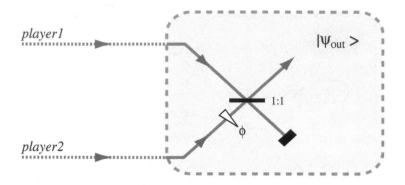

Figure 8.5 Schematic of the reconstruction protocol for {1,2}. ψ_{out}: reconstructed quantum state; 1:1: 50% reflectivity beamsplitter; ϕ phase delay.

8.4.2   Reconstruction Protocols

8.4.2.1   {1,2} Reconstruction Protocol

As proposed by Tyc and Sanders, the access structure formed by players 1 and 2, henceforth denoted as {1,2}, reconstructs the secret quantum state by completing a Mach–Zehnder interferometer, using a 1:1 beam splitter as shown in Figure 8.5. The resulting output from the interferometer can be expressed as

$$\hat{a}_{out} = (\hat{a}_1 + \hat{a}_2)/\sqrt{2} = \hat{a}_{in} \qquad (8.6)$$

Equation (8.6) clearly shows that the secret is perfectly reconstructed using the {1,2} reconstruction protocol. In contrast, more complex protocols are required for {2,3} or {1,3} to reconstruct the secret state.

8.4.2.2   {2,3} and {1,3} Reconstruction Protocol
Using Two OPAs

As the access structures {2,3} and {1,3} are symmetric, we will only study the {2,3} reconstruction protocol in the following. The {1,3} reconstruction proto-col follows an identical analysis. In the original proposal by Tyc and Sanders, the {2,3} reconstruction protocols require two additional optical parametric amplifiers, as shown in Figure 8.6. In this protocol, the access structure shares from {2,3} are interfered on a 1:1 beam splitter. The two output fields are then parametrically amplified and deamplified, respectively, using a pair of optical amplifiers, each acting on one of the outputs. The optical parametric amplifiers (OPAs) perform a unitary squeezing operation on the input fields, squeezing one of the quadratures while antisqueezing the orthogonal quadra-ture. The optical parametric amplifier acting on the first output squeezes the amplitude quadrature, while the second optical parametric amplifier squeezes the phase quadrature of the second output. The two beams are then

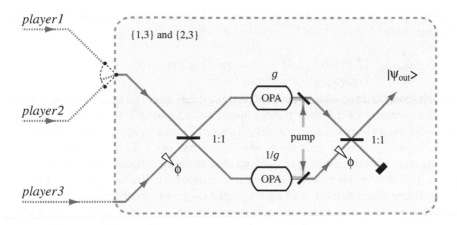

Figure 8.6 Schematic of the 2 OPA reconstruction protocol for {2,3} and {1,3}. OPA(g): optical parametric amplifier with parametric gain g; ψ_{out}: reconstructed quantum state; 1:1: 50% reflectivity beam splitter; ϕ phase delay.

interfered on another 1:1 beamsplitter. The resulting output beam can then be expressed as

$$\delta \hat{X}_{out}^{\pm} = \frac{1}{2\sqrt{2}} \left[\delta \hat{X}_{in}^{\pm} \left(\sqrt{g} + \frac{1}{\sqrt{g}} \right) + \frac{1}{\sqrt{2}} \alpha^{\pm} (\delta \hat{X}_{sqz1}^{\mp} + \delta \mathcal{N}^{+}) \right.$$

$$\left. + \frac{1}{\sqrt{2}} \beta^{\pm} (\delta \hat{X}_{sqz2}^{\pm} + \delta \mathcal{N}^{-}) \right] \tag{8.7}$$

where g is the gain of the optical parametric amplifiers, and the parameters are defined as $\alpha^{\pm} = (-1 \mp \sqrt{2})/\sqrt{g} + \sqrt{g}(-1 \pm \sqrt{2})$ and $\beta^{\pm} = (-1 \pm \sqrt{2})/\sqrt{g} + \sqrt{g}(-1 \mp \sqrt{2})$. By choosing a correct gain for the optical parametric amplifiers, it is possible to reconstruct the secret state. Setting the gain to $\sqrt{g} = \sqrt{2} + 1$, the quadratures of the reconstructed secret simplify to

$$\delta \hat{X}_{out}^{+} = \delta \hat{X}_{\psi}^{+} - \sqrt{2} \delta \hat{X}_{sqz2}^{+} \tag{8.8}$$

$$\delta \hat{X}_{out}^{-} = \delta \hat{X}_{\psi}^{-} - \sqrt{2} \delta \hat{X}_{sqz1}^{+} \tag{8.9}$$

Equation (8.8) shows that in the ideal limit of perfect squeezing and with correct gain for the optical parametric amplifiers, the access structure can perfectly reconstruct the secret state.

The Tyc and Sanders [14] original scheme requires significant resources including a pair of entangled beams and two optical parametric amplifiers. Furthermore, the reconstruction protocol requires that the gain of the phase-sensitive amplifiers be controlled precisely and that they have a high nonlinearity. Each of these requirements is difficult to achieve experimentally. High nonlinearity can be achieved either in Q-switched or mode-locked setups, or by enhancing the optical intensities within optical resonators. These techniques, however, result in losses and reduced quantum efficiency. For these reasons an experimental demonstration of this reconstruction protocol would be extremely difficult to achieve using existing technology.

8.4.2.3 {2,3} and {1,3} Reconstruction Protocol Using a Feedforward Loop

An alternative reconstruction protocol for {2,3} and {1,3} that simplifies the orginal scheme by replacing the phase sensitive amplifiers with linear optics and electro-optic feedforward (Figure 8.7) was proposed by Lance et al. [19]. In this protocol the shares are interfered on a 2:1 beam splitter. The beam splitter reflectivity is chosen so that the magnitudes of the noise fluctuations (due to the entangled state and noise contributions) on each of the two shares are equal on the beam splitter output b. With a correctly chosen relative phase between the input beams, the noise fluctuations on the phase quadrature of b are cancelled and the phase quadrature of the secret state is reconstructed. The noise on the amplitude quadrature of b is increased by this process. This noise can be cancelled, however, since it is correlated with noise on the amplitude quadrature of c. By detecting c and feeding forward the resulting photocurrent to displace the noise b, it is possible to cancel this noise and simultaneously reconstruct the amplitude quadrature of the secret state. Typically, in feedforward schemes, the fluctuations are directly applied to the optical field using an electro-optic modulator. This method can be quite inefficient, resulting in high losses. A more efficient method is to apply the fluctuations to a separate intense beam (a strong local oscillator) and then interfere this beam with b on a highly reflective beam splitter, as shown in Figure 8.7.

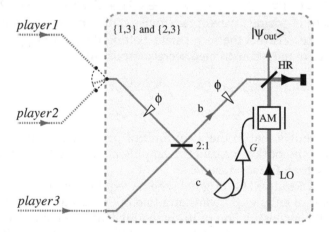

Figure 8.7 Schematic of the {2,3} and {1,3} reconstruction protocols for the (2,3) quantum state sharing scheme using linear optics and electro-optic feedforward. G: electronic gain; ψ_{out}: reconstructed quantim state; AM: amplitude modulator; LO: optical local oscillator; HR: high reflectivity; 2:1: 2/3 reflectivity beamsplitter; ϕ: phase delay.

The reconstructed secret using the feedforward protocol can then be expressed as

$$\delta \hat{X}_{out}^{+} = g^{+}\delta \hat{X}_{in}^{+} + \frac{\sqrt{3}}{2}\left(1 - \sqrt{3}g^{+}\right)\left(\delta \hat{X}_{sqz1}^{+} + \delta \hat{X}_{sqz2}^{+}\right)$$

$$+ \frac{1}{2}\left(g^{+} - \sqrt{3}\right)\left(\delta \hat{X}_{sqz1}^{-} - \delta \hat{X}_{sqz2}^{-}\right) + \left(\sqrt{3} - g^{+}\right)\delta N^{+} \quad (8.10)$$

$$\delta \hat{X}_{out}^{-} = \frac{1}{\sqrt{3}}\left(\delta \hat{X}_{in}^{-} + \delta \hat{X}_{sqz1}^{+} - \delta \hat{X}_{sqz2}^{+}\right) \quad (8.11)$$

where we have defined $g^{\pm} = \langle\hat{X}_{out}^{\pm}\rangle/\langle\hat{X}_{in}^{\pm}\rangle$ as the optical quadrature gains for the respective quadratures. The phase quadrature gain is constant $g^{-} = 1/\sqrt{3}$, while the amplitude quadrature gain can be controlled by varying the electronic feedforward gain G by $g^{+} = (1/\sqrt{3} + G/\sqrt{6})$. The specific gain of $g^{+}g^{-} = 1$ can be thought of as the *unitary gain point*, similar to the unity gain point in continuous variable teleportation experiments [17,18]. At this unitary gain point and in the ideal limit of perfect squeezing, the quadratures of the reconstructed state can be expressed as

$$\delta \hat{X}_{out}^{+} = \sqrt{3}\delta \hat{X}_{in}^{+}$$

$$\delta \hat{X}_{out}^{-} = \frac{1}{\sqrt{3}}\delta \hat{X}_{in}^{-}. \quad (8.12)$$

Hence this reconstruction protocol allows the access structure to reproduce a unitary transformed version of the secret state. Of course, ideally the reconstructed output state should be of identical form to the input state. This can be achieved here by performing a single unitary squeezing operation on the output state of Equation (8.12). Since the two OPAs reconstruction protocol requires two of these operations, the feedforward scheme even including this operation is significantly less demanding. It should be pointed out, however, that the result of Equation (8.12) is only possible if quantum resources (i.e., entanglement) are shared between the players in the protocol. The unitary transform required for the output state to be of the same form as the input state, on the other hand, requires only local resources and no entanglement. Therefore it is reasonable to conclude that the essence of the quantum state sharing reconstruction protocol is contained within the feedforward scheme rather than the unitary transform. For these reasons, we consider that the feedforward scheme in and of itself constitutes a {2,3} reconstruction protocol for quantum state sharing. It should also be noted that the squeezing exhibited on the feedforward reconstructed state is deterministically known. Therefore, if the quantum state sharing protocol was utilized within a quantum information network, more likely than not, the squeezing could be taken account of by simply adjusting the alphabet used by the network in subsequent processes.

8.4.3 Characterization

Quantum state sharing can be characterized in a similar manner to quantum teleportation and other quantum information protocols concerned with quantum state reconstruction. We characterize the quality of the state reconstruction using fidelity $\mathcal{F} = \langle \psi_{in} | \rho_{out} | \psi_{in} \rangle$, which measures the overlap between the secret and reconstructed quantum states [22]. While the secret state can, in general, be an arbitrary unknown state, we simplify the characterization by assuming that the secret is a coherent state. Since coherent states span Hilbert space, a demonstration of quantum state sharing with coherent states can be directly extended to arbitrary quantum states in general. Assuming that all fields involved have Gaussian statistics, the fidelity can be expressed in terms of experimentally measurable parameters as

$$\mathcal{F} = 2e^{-(k^+ + k^-)/4} \Big/ \sqrt{(1 + V_{out}^+)(1 + V_{out}^-)} \tag{8.13}$$

where we have defined $k^\pm = \langle X_{in}^\pm \rangle^2 (1 - g^\pm)^2 / (1 + V_{out}^\pm)$. The fidelity for the {1,2} reconstruction protocol can be determined directly. For the {2,3} and {1,3} reconstruction protocols, since the reconstructed state is a squeezed version of the secret state, the fidelity is determined by inferring the unitary parametric operation $\delta \hat{X}_{infer}^\pm = (\sqrt{3})^{\mp 1} \delta \hat{X}_{out}^\pm)$ on the reconstructed state. In the ideal case, $\delta \hat{X}_{infer}^\pm = \delta \hat{X}_{in}^\pm$. Any one of the access structures sets can, in the ideal case of perfect squeezing and at unitary gain, achieve perfect reconstruction of the secret quantum state $\mathcal{F} = 1$; the corresponding adversary structure obtains no information about the secret state $\mathcal{F} = 0$.

The efficacy of the quantum state sharing scheme can be characterized by determining the average fidelity over all access structure permutations. It is relatively easy to show that for a general (k, n) threshold quantum state sharing scheme without any entanglement resources, the maximum fidelity averaged over all access structure permutations is $\mathcal{F}_{avg}^{clas} = k/n$. This limit can only be exceeded by using quantum resources. For the $(2, 3)$ quantum state sharing scheme this limit reduces to $\mathcal{F}_{avg}^{clas} = 2/3$.

Quantum state sharing can also be characterized by measuring the signal transfer to $(\mathcal{T})$, and the additional noise on $(\mathcal{V})$, the reconstructed state [23]. This measure provides additional information to the fidelity measure about the efficacy of state reconstruction. Such analysis has been used to characterize quantum nondemolition [24] and quantum teleportation experiments [18]. Unlike the fidelity measure described above, both $\mathcal{T}$ and $\mathcal{V}$ are invariant to unitary parametric transformations of the reconstructed state. Therefore, for the $\mathcal{T}$ and $\mathcal{V}$ measures, it is unnecessary to infer a unitary transform after the {2,3} and {1,3} reconstruction protocols to characterize the state reconstruction.

The signal transfer describes the signal-to-noise transfer between the secret and reconstructed state for both quadratures $\mathcal{T} = T^+ + T^-$. It is expressed in terms of the quadrature signal transfer coefficients $T^\pm = SNR_{out}^\pm / SNR_{in}^\pm$, where SNR is the signal-to-noise ratio. The additional noise describes extra quadrature noise on the reconstructed state $\mathcal{V} = V_{cv}^+ V_{cv}^-$ and is expressed in terms of the conditional variances between the secret and reconstructed state

$V_{cv}^{\pm} = V_{out}^{\pm} - |\langle \delta \hat{X}_{in}^{+} \delta \hat{X}_{out}^{+} \rangle|^{2} / V_{out}^{\pm}$. $\mathcal{V}$ can be expressed in experimentally measurable parameters as $\mathcal{V} = (V_{out}^{+} - (g^{+})^{2})(V_{out}^{-} - (g^{-})^{2})$.

Any one of the access structures can, in the ideal case, achieve perfect state reconstruction corresponding to a signal transfer $\mathcal{T} = 2$ and additional noise $\mathcal{V} = 0$; the adversary structure obtains no information about the secret state $\mathcal{T} = 0$ and $\mathcal{V} = \infty$.

8.5 Experimental Realization

8.5.1 Experimental Setup

Quantum state sharing has recently been experimentally demonstrated by Lance et al. [25]. In this experiment, a Nd:YAG laser producing a coherent laser field at 1064 nm was used to provide a shared time frame or *universal local oscillator* among all parties. The secret quantum state was generated from this laser field, by displacing the sideband frequency vacuum states of the laser field using a phase and amplitude modulator.

The pair of quadrature entangled beams used in the dealer protocol were generated from the interference of two amplitude squeezed beams. The squeezed beams were produced using a pair of optical parametric amplifiers (OPAs) seeded with 1064 nm light and pumped with 532 nm light, produced from a second harmonic generator (SHG) [16]. A $\pi/2$ phase shift was chosen between the beams, which after interference on a 1:1 beam splitter exhibits quadrature entanglement.

In order to enhance the security of the secret state against the adversaries, the coherent quadrature amplitudes of the entangled beams were displaced with Gaussian noise. Experimentally, this can be achieved by encoding broadband Gaussian noise onto the nonlinear crystals within the OPA resonators at the sideband frequency of the secret state.

A homodyne detection system, consisting of a pair of balanced detectors and the universal local oscillator with controllable optical phase, was used to characterize the secret, adversary and reconstructed quantum states for the reconstruction protocols.

8.5.2 Experimental Results

Since the {1,3} and {2,3} reconstruction protocols are equivalent owing to the symmetry of the player 1 and 2 shares, the (2,3) threshold quantum state sharing scheme is demonstrated through the implementations of only the {1,2} and {2,3} reconstruction protocols.

Figure 8.8 shows the noise spectra for the secret and reconstructed state for the {1,2} protocol. The corresponding inferred Wigner function standard deviation contours were determined from these noise spectra and are shown in Figure 8.8(c). For the {1,2} protocol, the best measured fidelity was $\mathcal{F}_{\{1,2\}} = 0.93 \pm 0.02$ with corresponding optical quadrature gains of $g^{+} = 0.94 \pm 0.01$ and $g^{-} = 0.97 \pm 0.01$, respectively. Figure 8.8(d) shows several measured

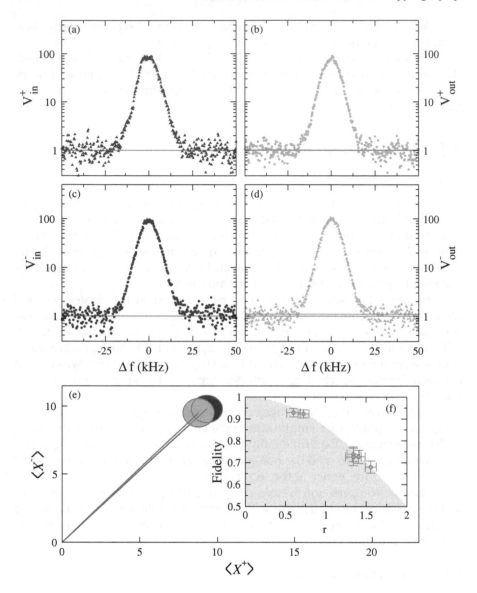

Figure 8.8 Experimental results for the {1,2} access structure. (a) Input amplitude quadrature; (b) output amplitude quadrature; (c) input phase quadrature; (d) output phase quadrature spectra of the secret quantum states. Δf is the offset from the signal frequency. Signal frequency $= 6.12$ MHz; resolution bandwidth $= 1$ kHz; video bandwidth $= 30$ Hz. (e) Standard deviation contours of Wigner functions of the secret (dark gray) and extracted (gray) quantum states. (d) Measured fidelity as a function of gain deviation $r^2 = ((\hat{X}^+_{out}) - (\hat{X}^+_{in}))^2 + ((\hat{X}^-_{out}) - (\hat{X}^-_{in}))^2$. Gray area highlights the accessible fidelity region. Points plotted are from six different experimental runs.

fidelity points for the {1,2} protocol as a function of the phase space distance, r, between the coherent amplitudes of the secret and reconstructed states. The nonzero distance of r on the experimental points is due to mode mismatch, optical losses, and imperfect phase locking. The corresponding adversary structure obtained a fidelity of $\mathcal{F}_{\{3\}} = 0$, since share {3} contains no information about the secret state.

Similarly, Figure 8.9 shows an example of the secret and reconstructed state for the {2,3} protocol. To allow for a direct measure of the overlap between the secret and reconstructed states, an inferred unitary squeezing operation was performed on the reconstructed state of this protocol. The inferred Wigner function standard deviation contour after this unitary squeezing operation is shown in the figure. Figure 8.10 shows the measured fidelity for a range of gains. The amplitude quadrature gain, and subsequently g^+g^-, was controlled by varying the gain G of the photocurrent of the electro-optic feedforward loop. At the unitary gain point, the best fidelity observed was $\mathcal{F}_{\{2,3\}} = 0.63 \pm 0.01$ with corresponding optical quadrature gains of $g^+g^- = (1.77 \pm 0.01)(0.58 \pm 0.01) = 1.02 \pm 0.01$ in this case. The corresponding adversary structure {1} achieved an average fidelity of $\mathcal{F}_{\{1\}} = 0.03 \pm 0.01$.

The quantum nature of the (2,3) threshold quantum state sharing scheme is demonstrated by the average fidelity over all the access structure permutations of $\mathcal{F}_{avg} = 0.74 \pm 0.04$, which exceeds the classical limit $\mathcal{F}_{avg}^{clas} = 2/3$. This can only be achieved using quantum resources and provides a direct verification of the tripartite continuous variable entanglement between the shares dealt to the players.

The quantum state sharing scheme was also characterized with the signal transfer (T) to, and the additional noise (V) on, the reconstructed state. The inset of Figure 8.11 shows the experimental T and V points obtained for the {1,2} protocol, plotted on orthogonal axes [23]. The theoretical point, assuming no losses, is also shown. The {1,2} protocol achieved a best state reconstruction of $T_{\{1,2\}} = 1.77 \pm 0.05$ and $V_{\{1,2\}} = 0.01 \pm 0.01$. Both of these values are close to optimal, being degraded only by optical losses and experimental inefficiencies.

Figure 8.11 shows the experimental T and V points obtained for the {2,3} protocol for a range of gains together with the theoretical curve for varying electronic feedforward gain G. The adversary structure {1} is also shown. The accessible region for the {2,3} protocol without entanglement is illustrated by the shaded region. The quantum nature of the state reconstruction is demonstrated by the experimental points that exceed this classical region. For the {2,3} protocol the lowest reconstruction noise measured was $V_{\{2,3\}} = 0.46 \pm 0.04$ and the largest signal transfer was $T_{\{2,3\}} = 1.03 \pm 0.03$. The measured experimental points with $T > 1$ exceeded the information cloning limit [18], demonstrating that for these points, the {2,3} has better access to information encoded on the secret state than any other parties. The adversary structure, meanwhile, obtains significantly less information about the state reconstruction, with a mean signal transfer and reconstruction noise of $T_{\{1\}} = 0.41 \pm 0.01$ and $V_{\{1\}} = 3.70 \pm 0.06$, respectively. The separation of the adversary structure

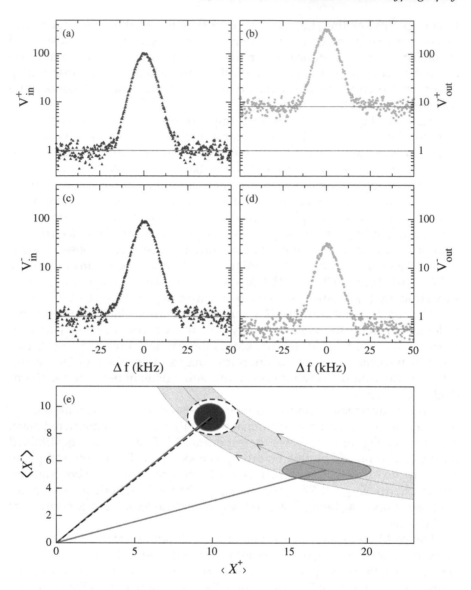

Figure 8.9 Experimental results for the {2,3} access structure. (a) Input amplitude quadrature; (b) output amplitude quadrature; (c) input phase quadrature; (d) output phase quadrature spectra of the secret quantum states. (e) Standard deviation contours of Wigner functions of the secret (dark gray) and reconstructed (gray) quantum states. The dashed circle represents the inferred quantum state $\delta \hat{X}_{\text{infer}}^{\pm} = (\sqrt{3})^{\mp 1} \delta \hat{X}_{\text{out}}^{\pm}$ after a local unitary operation (parametric amplification) is performed on the reconstructed state.

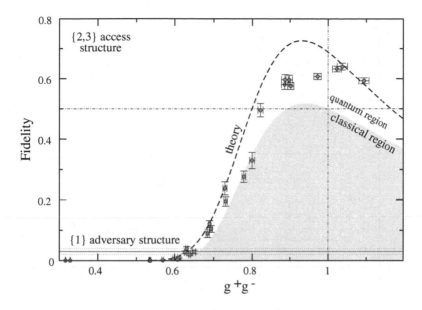

Figure 8.10 Experimental fidelity for the {2,3} access structure as a function of the product of the optical quadrature amplitude gains g^+g^-. Dashed line: calculated theoretical curve with squeezing of -4.5 dB, added noise of $+3.5$ dB, electronic noise of -13 dB with respect to the quantum noise limit, and feedforward detector efficiency of 0.93. Solid line and dotted lines: experimental fidelity for the adversary structure and error bar. Gray area highlights the classical boundary for the access structure.

T and V points from that of the {2,3} protocol in Figure 8.11 illustrates that in such a protocol the access structure performs far better than any adversary structure.

8.6   Applications of Quantum State Sharing

Quantum state sharing has many possible applications in quantum information science. We conclude this chapter with a discussion of some of those applications.

8.6.1   Quantum Information Networks

One of the primary experimental goals of quantum information science is to realize quantum networks analogous to the Internet. These *quantum information networks* are expected to consist of atomic nodes where quantum states are stored and processed, connected by optical channels. To date, research into quantum information networks has, for the most part, been restricted

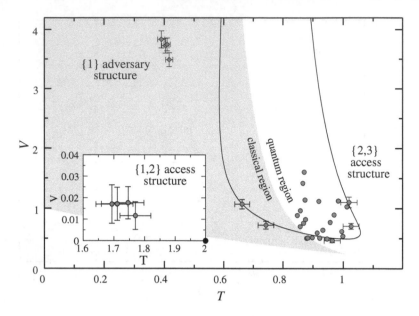

Figure 8.11 Experimental signal transfer (T) and additional reconstruction noise (V) for the {2,3} access structure for varying gain, and the adversary structure. Solid line: calculated theoretical curve with squeezing of −4.5 dB, added noise of +3.5 dB, electronic noise of -13 dB with respect to the quantum noise limit, and feedforward detector efficiency of 0.93. Gray area: the classical boundary for the {2,3} access structure. (Inset) Experimental T and V for the {1,2} access structure (gray points) and the theoretical point (black point).

to demonstrations of the individual components required for their success, such as quantum memory [26], and quantum gates [27]. Apart from achieving these components in isolation, it is essential that quantum information networks are scalable, both in terms of the complexity of problems solvable within a node, and in terms of the total number of nodes involved.

Quantum state sharing allows many nodes to cooperate on a specific computing problem as with distributed computing, and it naturally extends the number of nodes that can be involved in any quantum information protocol. Suppose, as an example, that a group of people who do not trust each other want to crack a code using Shor's algorithm [28]. Suppose also that the group has only one quantum computer powerful enough to realize this algorithm. They can ask the person owning this quantum computer to perform most, but not all, of the steps in Shor's algorithm and to use the quantum state sharing protocol to distribute the partial results to the other members of the group. Finally, an access structure can reunite, and with a smaller, less powerful quantum computer complete Shor's algorithm to crack the code.

These properties, coupled with its ability to facilitate quantum error correction, as discussed below, suggest that quantum state sharing is an important tool for scalable quantum information networks.

8.6.2   Quantum Error Correction

Error correction capabilities are required for any form of large-scale computation. In conventional computing, error correction is performed by introducing redundancy within the computer. The simplest form of redundancy is to encode each bit of information more than once. If an error or erasure occurs, it appears as a discrepancy among the multiple copies of the bit and can be corrected for. In quantum computing, error correction is made much more complex by our inability to clone perfectly, or nondestructively measure, a quantum state. Many techniques have been proposed to overcome these limitations for various forms of errors [29], and quantum state sharing is one example [9]. In the case of continuous variable quantum state sharing, error correction is possible for errors that can be seen to have occurred. One example of such an error is a faulty connection to one node in a quantum information network, resulting in the irreversible but known destruction of a quantum state under transmission to that node. For quantum state sharing, the redundancy required for successful error correction arises from the distribution of shares, each containing some fraction of the secret state, to n nodes (players). From the previous discussion we know that any k nodes (or nodes in a quantum information network) can collaboratively recover the secret state. Therefore, as many as $n - k$ nodes can malfunction with the secret state still perfectly retrievable.

8.6.3   Transmission of Entanglement
over Faulty Channels

One of the primary requirements of long-distance quantum communication and quantum information networks is the effective transmission of entanglement in nonideal environments. It has been shown that highly inefficient transmission lines can be overcome using entanglement purification, quantum memory, and quantum repeaters analogous to conventional repeaters in fiber-optic communication [30]. Such techniques, however, are of limited use if the communication channels, or network nodes, are prone to catastrophic failure. These sorts of failures commonly occur even in the present day Internet, for example when servers go off-line or are overloaded. It is therefore highly likely that catastrophic failures will also be exhibited by quantum information networks. Quantum state sharing overcomes these failures using a series of error correction protocols such as those described above. Let us consider the case of {2,3} secret sharing, with the goal of distributing one of a pair of entangled beams through a channel prone to catastrophic failures. The entangled beam is encoded into three shares, and the first is sent down the optical channel. If the transmission fails, the sender uses the other two shares to recover the entangled state and begins again. If the transmission is successful, the sender sends the second share. If transmission of the second share is successful the receiver can recover the entangled beam, and the entangled state transmission was successful. If the second transmission fails, the sender encodes the third and final share into three parts (subshares) and repeats the procedure. If the first two subshares are transmitted successfully the receiver can recover the third share and, since he or she already has the first share, also the entire entangled beam. If transmission of the second subshare fails, the sender encodes the remaining subshare into three subsubshares and the process continues. This nested series of quantum state sharing protocols provides in principle a 100% successful method to distribute entangled states, and indeed any arbitrary quantum states, through channels prone to catastrophic failures.

8.6.4   Multipartite Quantum Cryptography

Quantum state sharing could be useful for generalized quantum key distribution. In the usual quantum cryptography protocols, such as BB84 [31], Alice and Bob share a quantum key, which can be used to create a secure one-time pad via public channels. A quantum key can be established by Alice sending qubits to Bob, as suggested by Bennett and Brassard, or equivalently by sharing entangled pairs, or ebits, along the lines suggested by Ekert. To see how quantum state sharing plays a role in quantum key distribution, let us consider the latter approach of having Alice and Bob share ebits.

Suppose that Alice and Bob are not planning to use the key themselves and instead disseminate their shares to other players. Alice's colleagues have n_A shares, and Bob's colleagues hold n_B shares. At some future time, some of Alice's colleagues and some of Bob's colleagues can collaborate separately to extract Alice's state and Bob's state, respectively, and then communicate to establish the quantum key for secure quantum cryptography.

Acknowledgments

The authors would like to acknowledge Dr. Tomáš Tyc for his early contribution to the realization of the quantum state sharing scheme and Dr. Roman Schnabel for taking part in building the EPR source. We also acknowledge the Australian Research Council (ARC), the Defence Signals Directorate (DSD), the Defence Science and Technology Organisation (DSTO) and Alberta's informatics Circle Of Research Excellence (iCORE) for funding this project.

References

1. A. Shamir, How to share a secret, *Commun. ACM*, 22, 612–613, 1979.
2. M. Hillery, V. Buzek, and A. Berthiaume, Quantum secret sharing, *Phys. Rev. A*, 59, 1829–1834, 1999.
3. C.L. Liu, Introduction to Combinatorial Mathematics, McGraw–Hill, New York, 1968.
4. G.R. Blakley, Safeguarding cryptographic keys, *Proc. AFIPS* 1979, National Computer Conference, 313–317, AFIPS, 1979.

5. A. Herzberg, S. Jarecki, H. Krawczyk, and M. Yung, Proactive secret sharing, or how to cope with perpetual leakage, D. Coppersmith, ed., *Advances in Cryptology (CRYPTO95)*, Lecture Notes in Computer Science, No. 963, 339–352, Springer-Verlag, Berlin, 1995.

6. B. Schoenmakers, A simple publicly verifiable secret sharing scheme and its application to electronic voting, *Advances in Cryptology (CRYPTO99)*, Lecture Notes in Computer Sciences, Vol. 1666, 148–164, Springer-Verlag, Berlin, 1999.

7. National Institute of Standards and Technology (NIST), NIST FIPS PUB 185, Escrowed Encryption Standard, 1994.

8. W.K. Wootters and W.H. Zurek, A single quantum cannot be cloned, *Nature*, 299, 802–803, 1982.

9. R. Cleve, D. Gottesman, and H.K. Lo, How to share a secret, *Phys. Rev. Lett.*, 83, 648–651, 1999.

10. A. Karlsson, M. Koashi, and N. Imoto, Quantum entanglement for secret sharing and secret splitting, *Phys. Rev. A*, 59, 162–168, 1999.

11. W. Tittel, H. Zbinden, and N. Gisin, Experimental demonstration of quantum secret sharing, *Phys. Rev. A*, 63, 042301, 2001.

12. S. Wiesner, Conjugate coding, *Sigact News*, 15, 78–88, 1983; original manuscript written circa 1969.

13. S.D. Bartlett, H. de Guise, and B.C. Sanders, Quantum encoding in spin systems and harmonic oscillators, *Phys. Rev. A*, 65, 052316, 2002.

14. T. Tyc and B.C. Sanders, How to share a continuous-variable quantum secret by optical interferometry, *Phys. Rev. A*, 65, 42310, 2002.

15. Z.Y. Ou, S.F. Pereira, H.J. Kimble, and K.C. Peng, Realization of the Einstein–Podolsky–Rosen paradox for continuous variables, *Phys. Rev. Lett.*, 68, 3663–3666, 1992.

16. W.P. Bowen, R. Schnabel, P.K. Lam, and T.C. Ralph, Experimental investigation of criteria for continuous variable entanglement, *Phys. Rev. Lett.*, 90, 043601, 2003.

17. A. Furusawa, J.L. S¿rensen, S.L. Braunstein, C.A. Fuchs, H.J. Kimble, and E.S. Polzik, Unconditional quantum teleportation, *Science*, 282, 706–709, 1998.

18. W.P. Bowen, N. Treps, B.C. Buchler, R. Schnabel, T.C. Ralph, H.A. Bachor, T. Symul, and P.K. Lam, Experimental investigation of continuous-variable quantum teleportation, *Phys. Rev. A*, 67, 032302, 2003.

19. A.M. Lance, T. Symul, W.P. Bowen, T. Tyc, B.C. Sanders, and P.K. Lam, Continuous variables (2,3) threshold quantum secret sharing schemes, *New J. Phys.*, 5, 4.1–4.13, 2003.

20. T. Tyc, D.J. Rowe and B.C. Sanders, Efficient sharing of a continuous-variable quantum secret, *J. Phys. A: Math. Gen.*, 36, 7625–7637, 2003.

21. A. Einstein, B. Podolsky and N. Rosen, Can quantum description of physical reality be considered complete?, *Phys. Rev.*, 47, 777–780, 1935.

22. B. Schumacher, Quantum coding, *Phys. Rev. A*, 51, 2738–2747, 1995.

23. T.C. Ralph and P.K. Lam, Teleportation with bright squeezed light, *Phys. Rev. Lett.*, 81, 5668–5671, 1998.

24. J.Ph. Poizat, J.F. Roch, and P. Grangier, Characterization of quantum non-demolition measurements in optics, *Ann. Phys.* (Paris), 19, 265, 1994.

25. A.M. Lance, T. Symul, W.P. Bowen, B.C. Sanders, P.K. Lam, Sharing a secret quantum state, www.arXiv.org, quant-ph/0311015, 2003.

26. M.D. Lukin, Trapping and manipulating photon states in atomic ensembles, *Rev. Mod. Phys.*, 75, 457, 2003.

27. D.P. DiVincenzo, Quantum computation, *Science*, 270, 255–261, 1995.
28. P.W. Shor, Algorithms for quantum computation: discrete logarithms and factoring, *Proc. IEEE 35th Annual Symposium on Foundations of Computer Science*, 124, 1994.
29. P.W. Shor, Scheme for reducing decoherence in quantum computer memory, *Phys. Rev. A*, 52, R2493–2496, 1995.
30. L.M. Duan, M. Lukin, J.I. Cirac, and P. Zoller, Long-distance quantum communication with atomic ensembles and linear optics, *Nature*, 414, 413, 2001.
31. C.H. Bennett and G. Brassard, Quantum cryptography: public key distribution and coin tossing, in *Proc. IEEE International Conference on Computers, Systems and Signal Processing, Bangalore, India*, 175–179, 1984.

chapter 9

Free-Space Quantum Cryptography

C. Kurtsiefer
Ludwig Maximilians University Munich and
National University of Singapore

M. Halder
Ludwig Maximilians University Munich and
University of Geneva

H. Weinfurter and P. Zarda
Ludwig Maximilians University Munich and
Max-Planck-Institute for Quantum Optics

P.R. Tapster and P.M. Gorman
QinetiQ

J.G. Rarity
University of Bristol

Contents

9.1   Introduction...188
9.2   Quantum Coding 189
9.3   Quantum Cryptography (Key Sharing)............................190
    9.3.1   Faint Pulse Quantum Cryptography 190
        9.3.1.1   The Method 190
        9.3.1.2   The Tools 191
    9.3.2   The Long-Range Trial 195
    9.3.3   Entangled State Key Exchange 198
9.4   Space Applications 199
    9.4.1   Key Upload to Satellites 199
    9.4.2   Global Key Distribution via LEO Satellite...................200

    9.4.3   Satellite-to-Ground Quantum Key Distribution
           Utilizing Entangled Photon Pairs 200
    9.4.4   Other Quantum Key Exchange Scenarios 200
9.5   Experimental Feasibility of Key Exchange to Space 201
    9.5.1   Link Budgets for the Various Systems........................201
    9.5.2   Feasibility of Faint Pulse Quantum Key
           Distribution Systems 204
    9.5.3   Entangled State Quantum Cryptography: Feasibility 206
9.6   Conclusions 207
Acknowledgments 208
References 208

Abstract

This chapter describes the development of free-space quantum cryptography apparatus used for the secure exchange of keys. Existing systems use weak laser pulses to approximate single photons and polarization coding. Miniature multilaser sources and compact receiver units have been developed. These can be incorporated with lightweight portable telescopes to exchange cryptographic key material over long free-space ranges. The record distance to date has been 23.4 km between two mountain locations. Future experiments should be able to exchange keys over a 150 km range, and the feasibility of key exchange to a low Earth orbit satellite has been proven.

9.1   Introduction

With the exponential expansion of electronic commerce, the need for global protection of data is paramount. Data are normally protected by encoding them bit-wise using a large random binary number known as a key. An identical key is used to decode the data at the receiver. The secure distribution of these keys thus becomes essential to secure communications and transactions across the globe. At present electronic commerce generally exchanges keys using public key methods [1]. These methods rely on computational complexity, in particular the difficulty of factoring very large (publicly declared) numbers, as proof against tampering and eavesdropping. Any confidential information exchanged using such a key thus becomes insecure after a time when the rapid improvements in computational power or algorithmic development render the public key insecure. To guarantee long-term security, the cryptographic key must be exchanged in an absolutely secure way. The conventional method used for this for most of the last century has been the trusted courier carrying a long random key from one location to the other. Following the idea of Bennett and Brassard in 1984 [2], it is only recently that absolutely secure key exchange between two sites has been demonstrated over fiber [3–6] and free-space [7–13] optical links. This technique, known as quantum cryptography, has security based on the laws of nature and is, in principle, absolutely secure against any computational improvements. In this chapter we review the state of the art in free-space quantum cryptography. We describe a semiportable free-space quantum cryptography system that has been tested in a key exchange experiment between two mountain tops, Karwendelspitze (2244 m) and Zugspitze (2960 m), in southern Germany [12]. The distance between the two locations is 23.4 km. The elevated beam path dramatically reduced the air turbulence effects experienced in previous low-altitude tests [11] but also caused unprecedented requirements on stability against temperature changes, reliability under extreme weather conditions, and ease of alignment. In future high-altitude experiments we plan to extend this range more than 100 kilometers.

We go on to describe how such a system combined with sophisticated automatic pointing and tracking hardware could exchange keys with low Earth orbit satellites. If we engineer a satellite to be a secure relay station, we may see secure key exchange between any two arbitrary locations on the globe. The advantage of the space environment for communications is the loss-free (and distortion-free) optical path provided by the vacuum. Conventional optical free-space laser communication systems have been under development for some time. The recent success of the ARTEMIS-SPOT4 satellite-to-satellite (GEO-to-LEO) link [14] has increased confidence in these technologies. The question remains whether one can exchange a key to a low earth orbit satellite. Preliminary studies suggest this will be possible [15,16] with lightweight launch optics of ~ 125 mm aperture. In this chapter we discuss some of the detailed designs for such a system and remaining technical challenges to be overcome.

We also extend the scope of our study to introduce entangled state key exchange methods [17–22]. Such systems are intrinsically more secure than the faint pulse techniques that have predominated to date.

9.2 Quantum Coding

In quantum communications, the primary carrier of the information is the particle of light, the photon. The general qubit is represented by

$$|\Psi> = \alpha|0> + \beta|1> \tag{9.1}$$

with probability amplitudes normalized to $|\alpha|^2 + |\beta|^2 = 1$. The implicit assumption is that a single two-state system is involved. This generic notation can stand for any of the properties of various two-state systems, for example for ground |g> and excited |e> state of an atom, for horizontal |H> and vertical |V> polarization of a photon, or for path 0 and path 1 around an interferometer. The probability of detection of either state is the square modulus of the state amplitudes $|\alpha|^2$ and $|\beta|^2$. The key to quantum communications is the principle of superposition, where the probability amplitudes are both nonzero; the photon then exhibits wavelike and particlelike properties.

Another key concept for quantum communications is the phenomenon of entanglement. Entanglement describes the strong correlations that can exist in the quantum properties of two or more particles.

$$|\Psi\rangle = \frac{1}{\sqrt{2}}(|0\rangle_1|0\rangle_2 + |1\rangle_1|1\rangle_2) \tag{9.2}$$

is an example of a maximally entangled two-particle state. If one only looks at one of the two particles, one finds it with equal probability in state |0> or in state |1>. The state shows classical two-particle correlations in that when we measure a 1 (0) in channel 1 this immediately implies a 1 (0) in channel 2. The quantum state also shows strong correlation for any arbitrary superposition. For instance, if we consider a polarization entangled state, |0> is |H> and |1> is |V>. Measurements in any polarization direction in channel 1 will be 100% correlated in channel 2.

9.3   Quantum Cryptography (Key Sharing)

Using the above coding, one could encode data on single photons, but communication would be prone to errors due to loss; loss implies lost photons and thus lost bits. More practical one photon per bit schemes revolve around cryptographic key sharing where correlated random bit strings are generated at separate locations. Only those photons that arrive are used to form the key. Quantum key sharing has been demonstrated using faint pulses to approximate the one-photon superposition states [3–13] and using the strong correlations inherent in the entangled state [17–22].

9.3.1   Faint Pulse Quantum Cryptography

9.3.1.1   The Method

We follow the first experimental realization [7], which is known as the BB84 protocol and was first described in [2]. In this protocol the transmitter (Alice) encodes a random binary number in weak pulses of light using one linear polarization to encode 1's and orthogonally polarized pulses to encode 0's. To prevent eavesdropping, the number of photons per pulse is limited to much less than unity (the actual attenuation is linked to the overall transmission and is usually chosen as 0.1 photons per pulse). Furthermore, the encoding basis is randomly changed by introducing a 45° polarization rotation on half the sent pulses. In the receiver (Bob), single-photon-counting detectors detect the pulses, converting the light to macroscopic electronic pulses. The two polarizations are separated in a polarizing beam splitter, and a 0 or 1 is recorded depending on the detected polarization. A random switch selects whether to measure on a 0° or 45° polarization basis. Owing to the initial weak pulse and the subsequent attenuation along the transmission line, only very few of the pulses sent result in detected photoevents at the receiver. A record of when the pulses are detected is kept, and at the end of the transmission the receiver uses a classical channel (e.g., a telephone line) to tell the sender which pulses arrived and on what basis they were measured. All lost pulses and all detected pulses measured on a different basis to the encoding basis are erased from the sender's record. Thus identical random keys are retained by sender and receiver. Any remaining differences (errors) signal the interception of an eavesdropper! If an eavesdropper measures the polarization of one pulse, that pulse, being a single photon, is destroyed and does not reach Bob and thus is not incorporated in the key. The eavesdropper could choose a basis, measure the pulses, and then reinject copies. However, this strategy has to fail because half the time the eavesdropper will have chosen the wrong measurement basis and the reinjected pulses will induce an error rate of 25%. Of course a certain level of error could be caused by imperfections in the equipment used, but in order to guarantee absolute security any error should be attributed to (partial) interception. Below a certain threshold the error can be corrected and potential knowledge of the key by any eavesdropper can be erased by privacy amplification protocols [23,24].

9.3.1.2 The Tools

Compared to the original experiment using polarization rotations performed by high-voltage Pockels cells, it is by far advantageous to use separate laser diodes for every polarization at the transmitter. An additional simplification of the equipment can be achieved by randomly splitting the light in the receiver between the analyzers for two bases by a nonpolarizing beam splitter. This allowed us to design a long-range free-space key exchange apparatus capable of exchanging keys over free-space ranges greater than 20 km, where diffraction/turbulence and absorption losses reach up to 20 dB.

Transmitter The transmitter (Figure 9.1) is designed around an 80-mm-diameter transmit telescope. A novel miniature source of polarization coded faint pulses approximating single photons is used. The task of this transmitter module in a BB84 kit is to launch a faint light pulse with one out of 4 linear polarizations into a quantum channel, containing an average photon number of approximately 0.1. This ensures that the information leakage to

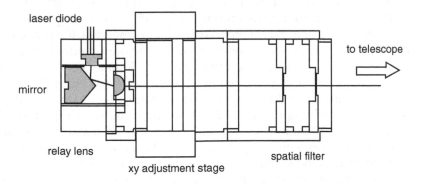

Figure 9.1 Optical configuration for a free-space transmitter module. The light of four laser diodes is combined with a mirror; here, only the diode emitting V polarized light is shown. Spatial indistinguishability is achieved by using a spatial filter. The module can be attached directly to a telescope.

an eavesdropper due to the possibility of the light field containing more than one photon is small [25].

A significant reduction in system complexity compared to former approaches was achieved by using four independent sources for the individual polarizations rather than optical elements selecting or preparing the polarizations as in earlier implementations of the BB84 protocol [7]. We have chosen laser diodes as light sources, because of the technological simplicity in achieving switching times of fractions of a nanosecond.

Another reduction of the necessary optical elements is enabled by the fact that light emitted by laser diodes shows a very high degree of polarization. Therefore not even passive optical polarization preparation elements like polarizers are needed. Together with an appropriate combination scheme for the light of the four laser diodes, the four necessary polarizations for the BB84 protocol are obtained by geometrical orientation of the laser diodes.

In principle, the combination of the light from the different sources could be achieved by beam splitters [26], but it turns out that a simple beam overlap in front of a spatial mode filter is sufficient [27]. A spatial filter is necessary to ensure that information is only encoded in the polarization degree of freedom and that the parasitic channel of emission direction is closed for the information encoded into the faint pulse.

Another important simplification arises from the fact that the laser diodes have to be attenuated strongly in order to come down to the 0.1 photon-per-pulse level on the quantum channel. This can be looked at as a very relaxed specification of the coupling efficiency into the quantum channel, which in our case was either a single-mode optical fiber or the spatial mode filter for a free-space optical link. Therefore, all the coupling into the quantum channel can be achieved using partial mode overlap between the laser diodes and the target mode of the optical fiber or the spatial filter.

All of this was elegantly achieved in a novel miniature source of polarization coded faint pulses (Figure 9.1). It consists of four laser diodes (850 nm wavelength) arranged on a ring around a conical mirror. Each laser is rotated to produce one of the four polarizations $0°$, $90°$, $45°$, or $135°$ and illuminates a spatial filter consisting of two pinholes with a diameter of 100 μm spaced at a distance of 9 mm. Since the overlap of the emission modes of the four laser diodes with the filter mode is rather poor, the initially very bright laser pulses are attenuated to about the required "one photon per pulse" level. The actual attenuation can be fine tuned by manipulating the diode current and precisely calibrated by optionally shining the light transmitting the spatial filter onto a single-photon detector. The filter erases all spatial information about which laser diode fired. Spectral information is also not attainable by an eavesdropper, as the spectra of the four laser diodes well overlap with a width of about 3 nm in pulsed mode. A continuous wave alignment laser was also fed through the spatial filter in order to ease optimizing the focusing of the receiver. The complete optical setup (Figure 9.2) was confined into an aluminum block of size $35 \times 35 \times 35$ mm to maintain rigid alignment, demonstrating the ability to integrate the source module, e.g., into a PC-based

Figure 9.2   Transmitter light source with spatial filter removed. The pulse driver with the computer interface is attached on the back side of the unit. The size is small compared to a telescope needed for focusing the beam to a distant location.

quantum cryptography system. The close thermal coupling of the laser diodes also ensures a fixed wavelength relation between the individual laser diodes, keeping the laser frequency side channel for a possible eavesdropping attack closed.

The beam from the filter transmitting only one spatial mode is transformed in a Galilean telescope to a collimated beam with a diameter of $\sim$ 40 mm FWHM (Figure 9.3). Together with the alignment laser and the single-photon detector, the whole system is mounted on a 25 × 50 cm breadboard, attached to a microradian-sensitive pointing stage on a sturdy tripod. The lasers are randomly driven from a computer via a digital output card at a 10 MHz repetition rate using subnanosecond duration pulses. This creates $\sim$ 500 ps duration optical pulses randomly polarized in 0°, 90°, 45°, or 135° directions. The computer uses a prestored random number to choose the polarization for the present set of experiments. Alternatively, nearly real-time generation was possible. A sequence of bits produced by a quantum random number generator running at 20 MHz could be fed to the Alice computer seconds before the transmission.

Receiver   The receiver system (Figure 9.4) consists of a 25-cm-diameter commercial telescope (Meade LX200) with computer controlled pointing capability. Unfortunately, the resolution of the mechanics of this system was the limiting factor for the alignment of the receiver and was also difficult to handle in the harsh outdoor conditions. Yet the stability of the system was very convincing.

A compact four-detector photon-counting module [27,28] was coupled to the back of the telescope after an RG780 long-pass filter to block out short-wavelength background. The module consists of a polarization-insensitive beam splitter passing two beams to polarizing beam splitters that

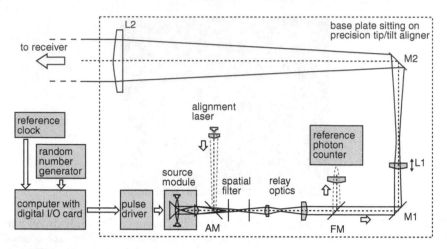

Figure 9.3 The Alice compact breadboard transmitter. The digital I/O card delivers a random 2-bit signal at 10 MHz synchronized to the reference clock. This signal is used in the pulse driver for randomly firing one of four lasers in the miniature source module. The four lasers are combined in a spatial filter using a conical mirror and relay lens. This system produces pulses with 0.05 to 0.5 photons per pulse. The output of the spatial filter is then transformed to a collimated beam with 2 mm FWHM and further expanded in a x20 telescope (L1 and L2) to produce a near diffraction-limited 40 mm beam. A precision translator with lens L1 allows for the fine focus adjustment. A bright CW laser beam can be injected with an auxiliary mirror AM for alignment purposes into the the same spatial filter as the faint pulses, while a calibration of the number of photons per bit can be made by inserting mirror FM and measuring a reference photocount. Mirrors AM, FM, M1 and M2 are gold coated for high reflectivity in the infrared.

are followed by four photon-counting avalanche diodes. One polarizing beam splitter is preceded by a 45° polarization rotator (half-wave plate). Photons detected in this channel are thus measured on the 45° basis, while the other polarizer allows measurement on the 0–90° basis. Since the splitting of incoming photons to the two analyzers by the beam splitter is truly random, no random number sequence is required on the receiver side, at the expense of more photodetectors.

The time of arrival of each photodetection is recorded in the computer using a two-channel time digitization card (Guide Technology GT654). Thus the four detectors' outputs are combined into the two channels with a delay of 5 ns. The delay is then used to discriminate between the two measurement bases. The overall optical detection efficiency of the receiver is about 16%, and the timing jitter was smaller than 1 ns.

Timing and synchronization The two separate computers were linked via modems operating over a standard mobile telephone link (9.6 Kbaud bit rate). Local oven-stabilized 10 MHz clocks were synchronized to better

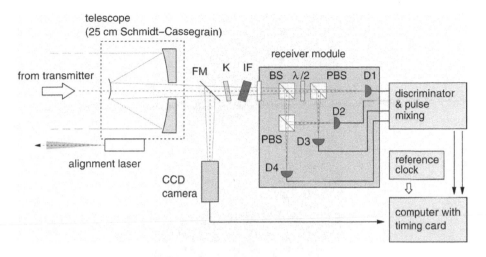

Figure 9.4 The receiver (Bob) consists of a 25 cm aperture Schmidt–Cassegrainian telescope. The miniature detector module is attached to the rear mounting of the telescope. It consists of a nonpolarizing beam splitter (BS) followed by two polarizing beam splitters (PBS). Single-photon detectors (D1-4) receive the output of the polarizers. In the D1/D3 arm, a half-wave plate rotates the analyzed polarization to the 45° basis. The module incorporated high-voltage supplies and discriminator circuitry to produce standard NIM pulses at the output. The detector outputs D3, D4 are combined with the D1, D2 outputs with a delay of 5 ns and input into the two-channel timing card in the PC. A flip mirror allows a CCD camera to view the incoming light for alignment purposes.

than 1 ns using a software-based phase-locked loop driven by the received photodetections. The photodetections thus can be gated in two 1.4-ns-wide time windows separated by 5 ns. Pulses outside these timing gates are ignored. The error rate due to dark and background counts is thus suppressed by a factor of ~ 1/35. The random polarization pulses are sent in 700 ms blocks preceded by a series of predetermined pseudorandom data sets lasting 110 ms to determine uniquely the start time of each block. Following the transmission of the block, a settling time of ~ 300 ms allows the computers to verify a successful transmission. Gross block length is thus just over 1.1 s. Sifting and error correction of the 700 ms data blocks were then performed over the telephone link using software developed in our 1.9 km experiment [11].

9.3.2 The Long-Range Trial

To avoid air turbulence effects, the long-range experiment was carried out over an elevated path with the receiver on Karwendelspitze and the transmitter located at a small experimental facility of the Max-Planck-Institute for Extraterrestrial Physics on the summit of Zugspitze, southern Germany. Initial alignment of both transmitter and receiver was achieved by shining a

Figure 9.5 The Alice breadboard is in the foreground. Karwendelspitze is in the background.

low-power green laser (3 mW) from Karwendelspitze that was clearly visible from Zugspitze. This was followed by a fine alignment using an approximately 500 μW beam at 850 nm passing through the same spatial filter as the faint light pulses in the Alice setup.

The collimation in the Alice system was adjusted to minimize the diameter at 23.4 km, resulting in a beam 1–2 meters in diameter (depending on air turbulence). This led to lumped optical losses in the transmission path of about 18 to 20 dB. With a receiver efficiency of around 16% and using faint pulses containing 0.1 photons per bit, the detected bit rate at Bob was about 1.5 to 2 kilobits per second.

Several night trials were carried out at various times from September 2001 to January 2002. In the final trial, several keys were exchanged over a period of three days at a selection of pulse intensities ranging from 0.4 to 0.08 photons per bit. Representative data are shown in Table 9.1. Total gated photon rates

Figure 9.6   The 25 cm receiver telescope with compact detector module attached.

(summed over all detectors) were 2–4 Kcps with dark counts dropping as low as 4 Kcps (actual detector dark counts summed to about 1 Kcps). Background rates were high because of scattered light from snow cover and the use of simple colored-glass shortwave blocking filters. The bracketed figures in the error rate column represent the errors arising from background counts alone, showing that half the error rate was from this source. The remaining errors of 2.1 to 2.7% arose from imperfections in the polarization encoding and decoding. At these error levels the error correction efficiency was between 50 and 60%. The low bit rate (9.8KB) of the mobile telephone link was a limiting factor in the sifting and error correction process. The final net key rate is about 1/6 of the raw detected rate. A factor of 2 is lost in sifting, a further factor of ~ 2 is lost in error correction, and then the data block efficiency is about 66%. Typically the sifting rate was 2 to 300 raw key bits per second (about 4 bytes of timing data per key bit). The interactive error correction process proceeded at a similar rate but of course with half as many key bits after sifting. To save time, longer blocks of raw key data were analyzed offline as shown. One point to note is that the error correction efficiency is better for these larger

Table 9.1   Summary of Selected Experiments

Night	Number of Photons Per Bit (+/−10%)	Unsifted Data Bits/s	Background Bits/s	Quantum Bit Error Rate %	Final Net Key Rate Bits/s	Total Key Exchanged Bits
16/01	0.37	4484	6268	4.11 (1.96)	626	9395
16/01	0.27	2505	5504	5.24 (3.08)	396	4341
16/01	0.18	2651	5578	4.54 (2.94)	363	5448
17/01	0.096	2627	4516	4.77 (2.41)	367	5399

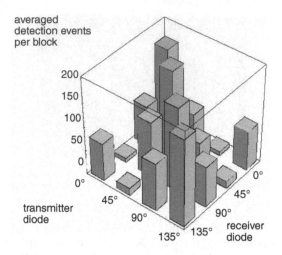

Figure 9.7 The detector matrix showing, on the diagonal, the number of bits transmitted and measured on the same basis. The suppressed measurements are the errors, bits encoded and measured on the same basis, but turning up in orthogonal channels (e.g., 0° and 90°). Measurements where different bases were used show roughly half the height of the diagonal, but constitute roughly half of all measurements.

keys because of the need to estimate the error rate sacrificing some of the key before correction.

Sifted and corrected keys showed a high degree of randomness with little or no bias. An online check of the balance of received bits compared to sent bits could be performed. A typical matrix of sent versus measured polarizations is shown in Figure 9.7. Care was taken to keep the diagonal values as close as possible.

9.3.3 Entangled State Key Exchange

Entangled state quantum cryptography is schematically illustrated in Figure 9.8. A source of (polarization) entangled photon pairs is configured to send one photon to Alice and one photon to Bob. Using receiver units identical to the one used in faint pulse cryptography (Figure 9.4) and selecting coincident detections measured on the same polarization basis, they establish identical random strings to use as cryptographic keys. There is no encoding of a random number to form the basis of the key, as the randomness comes from the superposition in the entangled state. This gives key security advantages over the faint pulse systems.

The first attempts at free-space entangled-state distribution have already begun. In [21], strong polarization correlations at 600 m range (across the river Danube) were demonstrated. These experiments could be easily extended to demonstrate entangled state key exchange [20]. As small telescopes were used, the lumped optical losses were in the 30 dB region. Pair photon

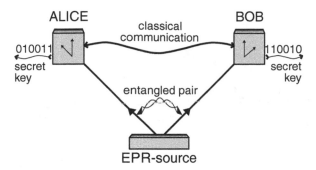

Figure 9.8 Schematic entangled pair key exchange system. Alice and Bob measure the arriving photons on 0 or 45 degree bases using receivers described in Figure 9.4. Keeping only those coincident detections measured on the same basis, they are able to establish identical keys.

coincidence rates of 2.10^4 per second were measured in a laboratory environment using only 18 mW of diode laser to pump a BBO nonlinear crystal. This led to long-range coincidence rates of 20 per second (or a raw key rate of 20 bits per second). Recently a full key exchange has been implemented over several kilometers [22].

In the next generation of portable equipments, pair brightness could be increased by at least an order of magnitude so that similar key rates to faint pulse systems could be achieved.

9.4 Space Applications

9.4.1 Key Upload to Satellites

Satellites storing and distributing large amounts of data such as digital TV transmitters and Earth observation satellites need to scramble data before transmission to licensed users on the ground. The scrambling and unscrambling process is done using symmetric key encoding data at the satellite and decoding them at the ground. New keys are thus required regularly to ensure the security of the transmissions. Key upload to satellites is thus a possible first application of quantum communications in space. The feasibility of key exchange to low Earth orbit is now being studied [15, 16]. The system will have to work between a range of 600 and 2000 km with fully automated acquisition and maintenance of the link during the satellite pass. Optical losses (see Section 9.5) in a typical pass could be as low as 20 dB for a close approach, rising to 35 dB at the longest range, which is around the maximum loss tolerance of a faint pulse system. Higher losses could be tolerated if one used true single-photon sources or accepted the reduced security associated with high average photon numbers per bit (< 0.5 photons/pulse).

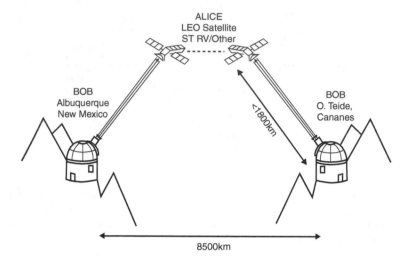

Figure 9.9   Global key exchange using quantum cryptography.

9.4.2   *Global Key Distribution via LEO Satellite*

It should be noted that a cryptographic terminal mounted on a satellite is just one side of a much larger experiment, resulting in a prototype global key exchange system. Since space vehicles in polar orbits can access the majority of the surface of the Earth, it makes sense to provide more than one ground receiving station. Keys exchanged on one continent can be used to secure those exchanged on another (Figure 9.9). The result is a truly strategic secure communications system.

9.4.3   *Satellite-to-Ground Quantum Key Distribution Utilizing Entangled Photon Pairs*

Using entangled (EPR) photons, it is possible to transmit secure keys to two places at once (Figure 9.10). The technology required for this system is still immature but could easily be integrated with that developed for the experiments described above. One key limitation may be the losses associated with the two optical paths from space to ground. Typically, optical losses due mainly to diffraction will be (at best) around 15 dB (see Section 9.6). The system will thus have to be able to cope with about 28–32 dB losses in the pair photon rate.

9.4.4   *Other Quantum Key Exchange Scenarios*

Further applications include satellite-to-satellite key exchange, particularly LEO to MEO and GEO satellites. Here the ranges are 1000 to 35,000 km. However the hardware will be similar to that developed for ground-to-satellite key exchange, albeit with higher losses. An attractive proposition is to have a satellite in GEO exchanging keys with a ground station or with LEO satellites

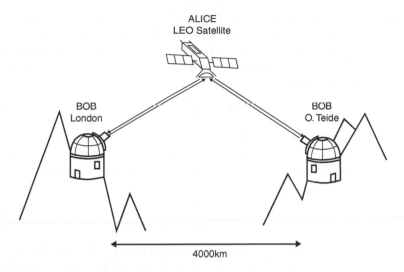

Figure 9.10 Satellite-to-ground quantum cryptography using entangled photons.

as in the ARTEMIS-SPOT4 classical optical communications experiment [14]. Such a satellite would be able to relay instantaneously keys between any two points on a hemisphere, and fixed pointing ground stations could be used. However, present technology on ARTEMIS produces a 7 µR divergence beam at a distance of 35,000 km. This then produces a ground spot 250 m in diameter on Earth. Even with a 2.5 m diameter telescope, diffraction losses will exceed 40 dB. Losses in a GEO-to-LEO link are much higher, as LEO launched telescopes would be limited in size to 25–30 cm in diameter.

9.5 Experimental Feasibility of Key Exchange to Space

9.5.1 Link Budgets for the Various Systems

Losses arise from various sources:

> Diffraction spreading of the beams
> Beam wandering due to atmospheric turbulence
> Pointing wander
> Atmospheric absorption (for ground-to-satellite experiments)
> Lumped receiver efficiency

Diffraction losses are set by the dimensions of the output telescope and by the standard equations of Gaussian optics. The divergence half-angle ($1/e^2$) for Gaussian beams with $1/e^2$ radius W_0 is given by

$$\theta = \frac{\lambda}{\pi . W_0} \tag{9.3}$$

In a standard collimating telescope the small-diameter beam is defocused by a negative lens and expands to fill the output lens, which collimates to near the diffraction limit. We can ensure that 98% of our Gaussian beam passes through this lens of diameter D when $2W_0 = 0.7D$. This sets the minimum divergence of our collimated beam to

$$2\theta = \frac{1.8\lambda}{D} \tag{9.4}$$

For instance the typical full divergence for a 125 mm lens illuminated by 650 nm light is thus $\sim 10\ \mu R$. At the receiver (range R) we intercept a large Gaussian beam diameter $2W = 2\theta R$; with small-diameter telescope D_T we will collect a fraction

$$L_g = 1 - \exp\left[\frac{2D_T^2}{(2W)^2}\right] \approx \frac{D_T^2}{2W^2} \tag{9.5}$$

Another source of beam broadening is the atmospheric turbulence. This causes beam wander and also scintillation. We estimate from our results at high altitude [10] that scintillation will cause effective beam wander of order 10–40 μR, depending on atmospheric conditions. Looking upwards from a high altitude ground station, this could be as low as 3–5 μR. Losses due to scattering from aerosols in the atmosphere are low at altitudes greater than 2000 m, being approximately 0.04–0.06 dB/km in clear weather. Obviously haze and cloud can increase these values significantly. Experiments at sea level will tend to have higher aerosol attenuation. Further simulation of atmospheric transmission using programs such as Modtran would improve these estimates.

We summarize the typical losses to be expected in various experimental scenarios in Table 9.2. For the comparison of systems we assume that satellite transmitter optics should be small and limit ourselves to 125 mm apertures where possible. This then gives a 10 μR diffraction spread (assuming light of around 650 nm wavelength), and we assume that pointing can be achieved to much better than this accuracy. In particular cases (GEO-to- ground and teleportation experiments) we allow for larger 300 mm apertures. This gives a smaller diffraction spread $\sim 4\ \mu R$ (we assume some improvement from existing classical experiments [14]), which is balanced against greater problems from pointing errors. The receiver optics need not be pointed to the diffraction limit as detectors can have a relatively wide field of view (up to 50 μR). However, space telescopes are still limited in dimension by weight. We limit our table to space receiver telescopes of aperture 300 mm in diameter. On the ground, tracking optical telescopes are available up to 1 m in diameter, while fixed telescopes up to 2 m in diameter might be used for ground-to-GEO systems.

Also included in the table is a next-generation high-altitude experiment where we might aim for 150 km key exchange using a faint pulse system as shown in Figure 9.3, Figure 9.4, and Figure 9.5. We also have options for an up-looking key exchange to LEO satellite Alice and a down-looking system with a ground Alice. For the entangled-state system we include losses from both arms and limit the satellite range to 700 km to limit loss. What is clear is

Table 9.2   Link Budgets for the Scenarios Described in the Text

Experiment	Range	Optics Alice 2θ	Optics Bob D_T	Diffract: Beam Spread	Scintillation Beam Spread	Atmospheric Atten dB	Lumped Loss dB
High altitude fixed stations	150 km	125 mm	a.25 cm b. 1 m	10 μR	10–40 μR	6–9 dB	a. 22–35 b. 16–29
Satellite Alice to Ground Bob	1000 km	125 mm	1 m	10 μR	0	2–5 dB	18–22
Ground Alice to Satellite Bob	1000 km	300 mm	300 mm	4 μR	3–5 μR	2–5 dB	21–27
GEO Alice to Ground Bob	36000 km	300 mm	2 m	4 μR	0	2–5 dB	36–41
Photon pairs QKD sat-ground	700 km + 700 km	125 mm	1 m	10 μR	0	2–5 dB	28–32

that for most systems in Table 9.2, the loss budget can be between 20 and 30 dB, and systems tolerating this sort of loss are well within our reach.

Receiver efficiency is common to all systems. Detector efficiencies of 70% are easily achieved for near-infrared wavelengths, while narrow-band filtering can also reach 70% transmission. By optimally coating telescope mirrors, transmission in the near-infrared can reach 90%, while relay lenses, beam splitters, and waveplates in the Bob module can reduce efficiency by 60%. All in all, a good target is to design a receiver with efficiency of about 30% ($\eta = 0.3$).

9.5.2 Feasibility of Faint Pulse Quantum Key Distribution Systems

The system closest to realization is the faint pulse quantum key distribution system described in Section 9.3. Key factors defining the feasibility of

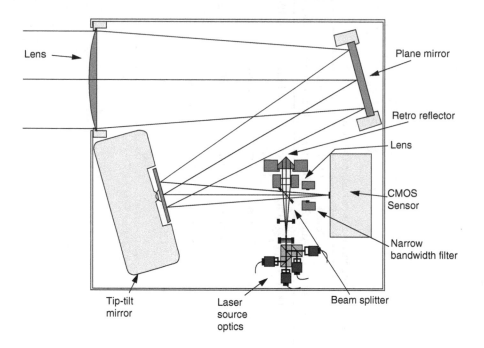

Figure 9.11   Breadboard for Alice suitable for development into a lightweight system for space operation. Includes tip–tilt mirror for closed-loop fine pointing and vibration compensation. The design is based on an f/10 telescope with 1 m focal length. The CMOS sensor (it could be a CCD in a final design) would have pixel size typically 10 μm, which with interpolation techniques would allow pointing to better than 10 μR over a field of view of 10 mR (1000 pixels). The retrocube system is used to produce a weak spot from the outgoing beam on the camera. Point ahead problems due to Doppler shifts are solved by displacing the image of the ground station guidestar from the image of the laser source. Note also the unlabeled spatial filter and narrow-band filter between source and first beam splitter.

long-range experiments are the system susceptibility to loss and the security of the key exchange in a high-loss system. The key exchange rate (K) will be a product of the pulse repetition rate of the system R, the number of photons per pulse M (typically $M < 0.1$), the atmospheric transmission T, the geometric loss L_g, the detection system lumped efficiency η, and the protocol efficiency (50%).

$$K = RMTL_g\eta/2 \qquad (9.6)$$

At present, repetition rates of 10 MHz can be achieved, but next-generation systems will run at 100 MHz. Assuming we can use $M \sim 0.1$ and take Alice on the satellite scenario, then the worst-case loss is ~ 22 dB (see Table 9.2). We then expect a raw key rate after sifting of $K \sim 10,000$ bits per second. Error correction and privacy amplification protocols will reduce this figure (by about 50%) but still leave long keys when satellite passes take 50 to 100 seconds within the 1000 km range.

The scenario where Alice resides in the satellite is preferred because typically it will be lower loss and the mass of the 125 mm telescope is quite low. In Figure 9.8 we show a preliminary compact breadboard designed for use on a microsatellite. The system is aimed at installation in a three-axis stabilized microsatellite with total mass < 50 kg. Satellite pointing to better than 0.5° (9 mR) is expected, so the tip–tilt pointing system would have to operate over this field of view with an accuracy much better than 10 μR. Using existing microsatellite technology [29] and suitable low-cost launch facilities, such a system could cost less than $20 million to bring to operation.

The system range is limited primarily by the background counting rate at the receiver, which contributes to the bit-error rate. With a background count rate per second B, the gated background count probability per pulse per detector is

$$P_b = Bt \qquad (9.7)$$

where t is a time gate set by the synchronization between transmitter and receiver. There are four detectors, and half the background pulses will lead to errors (the other half fortuitously match the sent bit). The background error rate is thus

$$E = 0.02 + \frac{2Bt}{MTL_g\eta} \qquad (9.8)$$

where T is atmospheric transmission. We assume a 2% base error rate due to optical element imperfections. The error correction scheme is optimum for error rates $E < 0.07$; thus we see that the minimum transmission is simply related to the background count rate

$$TL_g \geq \frac{40Bt}{M\eta} \qquad (9.9)$$

Clearly the error rate is independent of the pulse repetition rate but highly dependent on the gating. Present systems achieve around 1 ns synchronization and at night with 10 nm bandwidth filters the background can be reduced to $B < 100$ counts per second per detector. Putting these values into

Equation 9.9 with $\eta \sim 0.3$ and $M \sim 0.1$, we see that the system can tolerate about 40 dB of loss. Such low background rates are only possible during night operation or with subnanometer-wide filters in daylight (pointing sunwards is still a problem [16]).

In faint pulse quantum cryptography, the security is susceptible to eavesdropping on multiphoton pulses. This involves an eavesdropper hijacking the weak pulse beam at the exit of the transmitter, selecting only those pulses with more than one photon and measuring in such a way as to get partial information on the key. The eavesdropper can then reinject pulses at the receiver, thus bypassing the channel transmission losses. To avoid this possibility requires that the average number of photons per pulse (M) be low. A conservative assumption takes the view that all errors and all multiphoton pulses will leak information to an eavesdropper. After error correction, this implied leakage can be removed by applying a privacy amplification routine to both keys [23,24]. This random binary matrix multiplication scrambles the key and reduces key length while reducing the number of bits known by a potential eavesdropper to zero. An extreme technology projection assumes the eavesdropper can store one photon of a two-photon pulse (until measurement bases are revealed) and send the other to the receiver without seeing any loss. This scenario can be guarded against by limiting $M \leq TL_g$, which would suggests that a system operating at $M \leq 0.1$ can tolerate only 10 dB of transmission loss. However, the required technologies are decades away at present. With present technologies the eavesdropper is limited to a strategy where the coding basis and bit value are uniquely determined from three-photon detection events in a standard receiver [11]. To do this without discovery, the rate of three-photon detection at the eavesdropper must be greater than the normal rate of detection of single photons at the receiver. This implies that setting $M^2 \leq 24TL_g$ is an adequate security level to guard against this attack. A system operating at $M \leq 0.1$ is secure against attack with up to 34 dB of losses. This may affect the use of GEO-to-ground systems (see Table 9.2). It is probably more realistic, however, to protect against intercept–resend eavesdropping as discussed above using passive and active monitoring of the viability of the free-space channel. By definition, the entire free-space channel is visible during the key exchange, and any eavesdropper must remain invisible to all wavelengths of the electromagnetic spectrum that can be used to monitor the security. A new protocol based on varying the intensity may also make higher M possible in future [30].

9.5.3 Entangled State Quantum Cryptography: Feasibility

For a space experiment one could use 30-cm optics to collimate the pair photon beams to about 4 µR divergence. Two ground stations with 1 m telescopes could be used separated by up to 1000 km. A dedicated satellite in orbit at about 500 km altitude could then be arranged to pass between the two ground stations at a range of ~ 700 km from each. Losses per arm would then be between 14 and 16 dB in clear weather with total lumped losses of 28 to 32 dB. Present experiments produce around 20,000 detected pairs per second in the laboratory using compact diode lasers with 18 mW of power. Future systems will easily achieve an order-of-magnitude improvement leading to ground-based coincidence rates $\sim$ 200 per second with an effective raw key rate of 100 per second. Errors due to background count rates up to 10,000 per second are negligible because the background coincidence rate is around 10^{-1} per second if gates of 1 ns can be used. The pointing requirements for such an experiment are quite challenging. Two separate closed loop pointing and tracking systems are required with an accuracy of better than 4 µR. Down to 7 µR diffraction and beam wander is the state of the art in GEO-to-LEO classical communications experiments [14]. However, the mass of typical optical pointing and tracking terminals ($\sim$ 25 kg) mean that minimum experiment plus satellite mass would be in the 100 to 200 kg range.

9.6 Conclusions

As an important step toward satellite-based quantum cryptography we have demonstrated a secure key exchange over a free-space distance of 23.4 km. Operation down to 0.08 photons per bit has been demonstrated with optical losses of about 18 dB. A large fraction of errors arose from background counts but was still below 6%. Improved performance including daylight operation is expected with improved spatial filtering at the receiver and a narrow bandpass filter set to the correct wavelength together with accurate temperature control of the transmitter lasers. The apparatus showed high stability with the ambient temperatures in these experiments ranging from +5°C to −25°C. The polarization preparation and analysis modules developed in this work were stable and required no adjustments over the whole temperature range. In fact this was quite a relief, as system alignment is not a very pleasant task at 4:00 A.M., −20°C, and 2960 m altitude. Extension of entangled state key exchange experiments from the laboratory out to similar multikilometer ranges is already underway.

We have looked at space applications of quantum key distribution. We identified two future experiments/applications that could be achieved in the coming years. The first is global key distribution using satellite-to-ground faint pulse quantum cryptography (Figure 9.4) with secure bit rates greater than 1000 per second. The second is simultaneous key generation between two ground stations using entangled state quantum cryptography (Figure 9.5) with key generation at bit rates greater than 100 per second.

Obviously the nearest to implementation is the faint pulse quantum cryptography scheme, since a low-cost (but still greater than $20 million) microsatellite experiment is possible. A key problem yet to be solved is the extreme pointing and tracking requirements, although these are being addressed in classical optical communications experiments. Further work on brighter and lighter sources of pair photons will bring entangled state systems to space readiness. Similar improvements to sources will be required before teleportation experiments are ready to move out of the laboratory. A further requirement for all experiments is time synchronization, to nanoseconds for key exchange and picoseconds for teleportation schemes.

Acknowledgments

This chapter presents work supported by the European Space Agency under ESTEC/Contract No. 16441/02/NL/SFe Quantum Communications in Space.

References

1. See, for instance, Simon Singh, The Code Book: The Science of Secrecy from Ancient Egypt to Quantum Cryptography, Anchor, 1999.
2. C.H. Bennett and G. Brassard, in Proceedings of the IEEE International Conference on Computers, Systems and Signal Processing, Bangalore, India, IEEE, New York, 1984, pp.175–179.
3. P.D. Townsend, J.G. Rarity, and P.R. Tapster, Single photon interference in a 10 km long optical fibre interferometer, Electron. Lett., 29, 634–635, 1993; Enhanced single photon fringe visibility in a 10 km long prototype quantum cryptography channel, Electron. Lett., 29, 1291–1293, 1993.
4. A. Muller, J. Breguet, and N. Gisin, Experimental demonstration of quantum cryptography using polarised photons in optical fibre over more than 1 km, Europhys. Lett., 23, 383–388, 1993.
5. G. Ribordy, J.-D. Gautier, N. Gisin, O. Guinnard, and H. Zbinden, Fast and user-friendly quantum key distribution, J. Mod. Opt., 47, 517–531, 2000.
6. N. Gisin, G. Ribordy, W. Tittel, and H. Zbinden, Rev. Mod. Phys., 74, 145, 2002.
7. C.H. Bennett, F. Bessette, G. Brassard, L. Salvail, and J. Smolin, Experimental quantum cryptography, J. Cryptology, 5, 3–28, 1992.
8. J.D. Franson and B.C. Jacobs, Electron. Lett., 31, 232–234, 1995.
9. W.T. Buttler R.J. Hughes, S.K. Lamoureaux, G.L. Morgan, J.E. Nordholt, and C.G. Peterson, Practical free-space quantum key distribution over 1 km, Phys. Rev. Lett., 81, 3283, 1998.
10. W.T. Buttler, R.J. Hughes, S.K. Lamoureaux, G.L. Morgan, J.E. Nordholt, and C.G. Peterson, Daylight quantum key distribution over 1.6 km, Phys. Rev. Lett., 84, 5652–5655, 2000.
11. J.G. Rarity, P.M. Gorman, and P.R. Tapster, Secure key exchange over a 1.9 km free-space range using quantum cryptography, Electron. Lett., 37, 512–514, 2001; see also J.G. Rarity, P.M. Gorman, and P.R. Tapster, J. Mod. Optics, 48, 1887, 2001.
12. C. Kurtsiefer, P. Zarda, M. Halder, H. Weinfurter, P. Gorman, P.R. Tapster, and J.G. Rarity, Nature, 419, 450, 2002.
13. R.J. Hughes, J.E. Nordholt, D. Derkacs, and C.G. Peterson, New J. Phys., 4, 43, 2002; http://www.njp.org.
14. T. Tolker-Nielsen and G. Oppenhauser, In-orbit test result of an operational optical intersatellite link between ARTEMIS and SPOT4, SILEX, SPIE, 4635, 1, 2002 .

15. J.G. Rarity, P.R. Tapster, P.M. Gorman, and P. Knight, New J. Physics, 4, 82, 2002; http://ww.njp.org.
16. J.E. Nordholt, R.J. Hughes, G.L. Morgan, C.G. Peterson, and C.C. Wipf, Proc. SPIE, 4635, 117, 2002.
17. A.K. Ekert, Phys. Rev. Lett., 67, 661–663, 1991.
18. A.K. Ekert, J.G. Rarity, P.R. Tapster, and G.M. Palma, Phys. Rev. Lett., 69, 1293–1296, 1992.
19. W. Tittel, J. Brendel, H. Zbinden, and N. Gisin, Phys. Rev. Lett., 81, 3563, 1998.
20. G. Weihs, T. Jennewein, C. Simon, H. Weinfurter, and A. Zeilinger, Phys. Rev. Lett., 81, 5039–5043, 1998; T. Jennewein, C. Simon, G. Weihs, H. Weinfurter, and A. Zeilinger, Phys. Rev. Lett., 84, 4729–4732, 2000.
21. M. Aspelmeyer, H.R. Böhm, T. Gyasto, T. Jennewein, R. Kaltenbaek, M. Lindenthal, G. Molina-Terriza, A. Poppe, K. Resch, M. Taraba, R. Ursin, P. Walther, and A. Zeilinger, Science, August 1, 2002.
22. C.-Z. Peng, T. Yang, X.-H. Bao, J. Zhang, X.-M. Jin, F.-Y. Feng, B. Yang, J. Yang, J. Yin, Q. Zhang, N. Li, B.L. Tian, and J.-W. Pan, Experimental free-space distribution of entangled-photon pairs over 13 km, Phys. Rev. Lett., 94, 150501, 2005.
23. C.H. Bennett, G. Brassard, and J.-M. Robert, Privacy amplification by public discussion, SIAM Journal on Computing, 17, 210–229, 1988.
24. G. Brassard and L. Salvail, Secret key reconciliation by public discussion, Adventures in Cryptology, EUROCRYPT93, Lecture Notes in Computer Science, No. 765, Springer-Verlag, New York 1994, pp. 410–423.
25. N. Lütkenhaus, Phys. Rev. A, 61, 0523041-4, 2000.
26. S. Chiangga, P. Zarda, T. Jennewein, and H. Weinfurter, Appl. Phys. B, 69, 389–393, 1999.
27. C. Kurtsiefer, M. Halder, P. Zarda, and H. Weinfurter, Miniature modules for quantum cryptography, to be published.
28. J.G.Rarity, P.C.M. Owens, and P.R. Tapster, Quantum random number generation and key sharing, J. Mod. Optics, 41, 2435–2444, 1994.
29. See, for instance, Surrey Satellites, http://www.sstl.co.uk
30. X.-B. Wang, quant-ph/0411047, and X.-B. Wang, quant-ph/0410075.

chapter 10

Noise-Immune Quantum Key Distribution

Z.D. Walton, A.V. Sergienko, B.E.A. Saleh,
and M.C. Teich
Boston University

Contents

10.1   Introduction 212
10.2   Noise-Immune Polarization-Coded Schemes........................212
         10.2.1   Round-Trip Noise-Immune Polarization-Coded QKD.....212
         10.2.2   One-Way Noise-Immune Polarization-Coded QKD 214
         10.2.3   Symmetric Noise-Immune Polarization-Coded QKD 215
10.3   Noise-Immune Time-Bin-Coded Schemes 216
         10.3.1   Round-Trip Noise-Immune Time-Bin-Coded QKD 216
         10.3.2   One-Way Noise-Immune Time-Bin-Coded QKD 217
         10.3.3   Symmetric Noise-Immune Time-Bin-Coded QKD........221
10.4   Discussion 223
References 223

Abstract

We review quantum key distribution schemes that are noise-immune (require no alignment). For both polarization and time-bin qubits, we present three noise-immune schemes: round-trip, one-way, and symmetric. In the round-trip schemes, the signal travels back and forth between the legitimate users (Alice and Bob); in the one-way schemes, the signal travels only from Alice to Bob; in the symmetric schemes, a central source sends signals to Alice and Bob. The primary benefit of the symmetric configuration is that both Alice and Bob may have passive setups (neither Alice nor Bob is required to make active changes for each run of the protocol). We show that all the schemes can be implemented with existing technology.

10.1   Introduction

Of all the capabilities afforded by quantum information science [1], quantum key distribution (QKD; for a review, see Reference [2]) currently shows the most promise for practical implementation. Accordingly, there has been a concerted effort to develop QKD schemes that mitigate the technical challenges associated with existing approaches. Among the successes in this effort are the development of noise-immune (alignment-free) schemes for polarization [3] and time-bin [4–7] qubits. A further advance is the development of a symmetric scheme for time-bin qubits in which neither Alice nor Bob is required to make active changes to their setups [8]. Here we use the term symmetric to describe QKD schemes in which a central source distributes some number of photons to both Alice and Bob, so that they share entanglement. This is in contrast to round-trip and one-way configurations, in which the photons move according to Bob→Alice→Bob, and Alice→Bob, respectively. Here we show that symmetry and noise-immunity can be combined in a single implementation, for both polarization and time-bin qubits. Beginning with polarization-coded QKD, we first present a round-trip scheme in which noise-immunity is achieved by sampling the channel birefringence twice (once on the way from Bob to Alice and once on the way back). Second, we show how Klyshko's "advanced wave interpretation" (AWI) [9] can be used to transform this round-trip scheme into a one-way scheme imbued with passive detection. Third, we apply the AWI again to obtain a symmetric noise-immune scheme in which both Alice and Bob have passive setups. We then repeat these three steps for time-bin-coded QKD. For each scheme, we present a feasible implementation that relies only on current technology.

10.2   Noise-Immune Polarization-Coded Schemes

10.2.1   Round-Trip Noise-Immune Polarization-Coded QKD

The left column of Figure 10.1 shows the space-time diagrams of three noise-immune polarization-coded QKD schemes. For polarization qubits, noise-immune means that the scheme is immune to channel birefringence. The first scheme [Figure 10.1(A)] requires a round trip and is active (both Alice and Bob are required to make changes to their respective setups). The scheme runs as follows. Bob randomly chooses between polarization states $|V\rangle$ and $|H\rangle + |V\rangle$ (here, and for the rest of this chapter, we suppress normalization factors) and sends a single photon in that state to Alice. Alice uses a Faraday mirror to reflect that single photon back, and she also sends along an auxiliary unpolarized photon. Alice encodes a single bit by controlling the time ordering of the two photons she sends to Bob. Bob then measures each photon in the basis associated with the state of the initial photon he sent. Without knowing which state Bob sent to Alice, Eve cannot deterministically learn

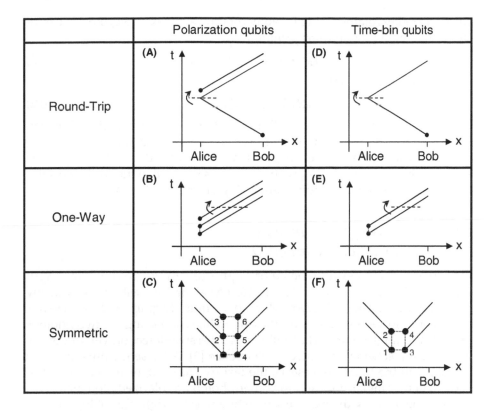

Figure 10.1   Space-time diagrams of six noise-immune QKD schemes organized by encoding (polarization or time-bin) and signal flow (round-trip, one-way, and symmetric). The dashed lines and curved arrows show how the advanced wave interpretation relates the round-trip schemes [(A) and (B)] to the one-way schemes [(B) and (C)], and the one-way schemes to the symmetric schemes [(C) and (F)]. The dotted lines connecting photons indicate entanglement. The photon labels in (C) and (F) are used later in this chapter.

Alice's bit setting. From Bob's point of view, the scheme is equivalent to Bennett's two-state protocol [10], since he is attempting to distinguish probabilistically between two nonorthogonal states. The noise-immune feature is derived from the unique property of the Faraday rotator: whatever the polarization transformation along the line from Bob to Alice, the photon that Alice reflects will arrive in Bob's laboratory in a polarization state orthogonal to its original state [11]. For example, if Bob sent $|V\rangle$, then either the first or the second photon he receives from Alice will be in the state $|H\rangle$. Thus if he measures one photon in state $|V\rangle$ and the other in state $|H\rangle$, he knows the value of Alice's bit. Any other detection pattern is ambiguous, and Alice and Bob discard these cases.

The AWI was originally conceived as a method for generating one-photon experiments from two-photon experiments. However, we may reverse this procedure and determine which two-photon state embodies the action of

Alice's Faraday rotator. Using Faraday rotation as an example, the AWI associates the single-photon transformation

$$H_{\text{in}} \rightarrow V_{\text{out}} \qquad V_{\text{in}} \rightarrow H_{\text{out}} \tag{10.1}$$

with the two-photon state

$$|H_{\text{in}} V_{\text{out}}\rangle + |V_{\text{in}} H_{\text{out}}\rangle. \tag{10.2}$$

In going from Equation (10.1) to Equation (10.2), the propagation direction for H_{in} and V_{in} is reversed. To preserve the handedness of the coordinate system, one of the transverse directions must be reversed as well. This may be accomplished by replacing V_{in} with $-V_{\text{in}}$. Thus we see that the AWI associates Faraday rotation with the polarization singlet state $|HV\rangle - |VH\rangle$.

10.2.2   One-Way Noise-Immune Polarization-Coded QKD

We arrive at the one-way scheme of Figure 10.1(B) by "folding" the input arm of the Faraday rotator of Figure 10.1(A) along the dashed line, thereby replacing a round-trip single-photon space-time diagram with a one-way, two-photon space-time diagram (the dotted line connecting the two photons indicates entanglement). What follows is a passive-detection version of the three-photon scheme presented in Reference [3]. Alice sends three photons to Bob, with the first two (case 1), the last two (case 2), or the first and last (case 3) in the singlet state and the other photon unpolarized. Bob makes his measurements using the passive setup shown on the right side of Figure 10.2. By appropriate postselection, this setup effectively makes a random choice of two out of the three photons and brings them together on a nonpolarizing beam splitter, which serves to distinguish the singlet state from the other three Bell states [12]. Ignoring the first Mach–Zehnder interferometer (with relative

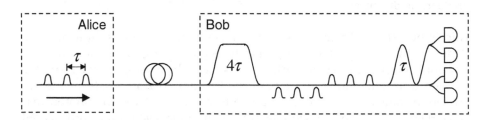

Figure 10.2   A schematic of one-way noise-immune polarization-coded QKD [see Figure 10.1(B)]. Alice sends three photons to Bob, with the first two (case 1), the last two (case 2), or the first and last (case 3) in the singlet state and the other photon unpolarized. The delay in Bob's first and second interferometer are 4τ and τ, respectively. Bob's apparatus effectively makes a random choice of two out of the three photons and brings them together on a nonpolarizing beam splitter, which serves to distinguish the singlet state from the other three Bell states [12]. The operation of the protocol is described in the text.

delay 4τ) for the moment, we see that the second interferometer (with relative delay τ) enables the first two, or the last two, photons to meet at the second beam splitter of this interferometer. If these two photons are in the singlet state, they will leave by opposite ports. The contrapositive is also true: if they leave by the same port (and are detected by one of the pairs of detectors on each output port), then one can infer that they were not in the singlet state. Returning to the first interferometer, we see that this interferometer provides an opportunity for the first and last photons to be analyzed in a similar way. Thus Bob's apparatus probabilistically chooses a pair out of the three photons sent by Alice and determines whether the pair is in the singlet state or in some orthogonal state [22]. Based on his detections, Bob can rule out at most one of the three cases corresponding to Alice's possible signal states. Therefore, after Bob has made his detection, Alice announces whether the run was a "data run" (cases 1 or 2), or a "test run" (case 3). The data runs are used to share key material, and the test runs are used to monitor the eavesdropper. The scheme is noise-immune because the singlet state is immune to collective rotation.

10.2.3 Symmetric Noise-Immune Polarization-Coded QKD

We can apply the AWI one more time to get a six-photon symmetric scheme [Figure 10.1(C)] from the three-photon one-way scheme by folding along the dotted line in Figure 10.1(B). As indicated in Figure 10.1(C), this would yield a six-photon entangled state. It is currently not practical to create such a state; however, we can still implement the scheme using three pairs of entangled photons in the state

$$|\Phi^+\rangle_{14}|\Phi^+\rangle_{25}|\Phi^+\rangle_{36}, \tag{10.3}$$

where $|\Phi^+\rangle = |HH\rangle + |VV\rangle$. The execution of the protocol is similar to the one-way polarization protocol, except that instead of randomly choosing a three-photon state and sending it to Bob, Alice uses the the apparatus depicted in Figure 10.3 to choose randomly which pair of photons is in the singlet state. For example, if Alice obtains a triple coincidence that indicates that photons

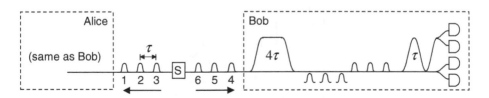

Figure 10.3 A schematic of symmetric noise-immune polarization-coded QKD [see Figure 10.1(C)]. A central source (S) emits three entangled pairs, so that Alice and Bob each get one from each pair. The scheme works much the same as the one-way scheme of Figure 10.2, except that Alice's apparatus makes a passive choice of the signal state that Bob receives, as described in the text.

1 and 2 were in the singlet state, then she knows that photons 4 and 5 are also in the singlet state. This effect can be seen as an application of entanglement swapping [13]. A similar argument works for the other two possible photon pairs on Alice's side. Thus, on these occasions, she effectively prepares for Bob one of the three signal states from the one-way scheme of Figure 10.1(B). The protocol then runs exactly as that of Figure 10.1(B). The security of the scheme derives from the fact that only a triple of maximally entangled photon pairs will produce the correlations that Alice and Bob measure. Therefore the source can be controlled by the adversary without compromising security.

10.3   Noise-Immune Time-Bin-Coded Schemes

10.3.1   Round-Trip Noise-Immune Time-Bin-Coded QKD

Figure 10.4 contains a schematic and space-time diagram of round-trip noise-immune time-bin-coded QKD (originally introduced as plug-and-play quantum cryptography [4]). The protocol begins with Bob launching a strong

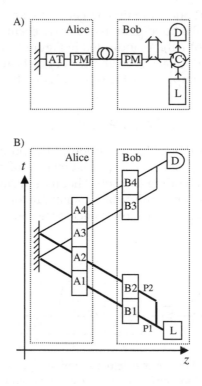

Figure 10.4   Schematic (A) and space-time diagram (B) for round-trip noise-immune time-bin-coded QKD. L is a source of laser pulses, C is a circulator, AT is an attenuator, and PM is a phase modulator.

pulse from a laser (L) into a Mach–Zehnder interferometer via a circulator (C). This interferometer splits the pulse into an advanced amplitude (P1) and a retarded amplitude (P2). The amplitudes travel through phase modulators (PM) on Bob's side and Alice's side, and are then attenuated (AT) to the single photon level and reflected by Alice back to Bob. Although both P1 and P2 will again be split at Bob's Mach–Zehnder interferometer, by gating his detector appropriately, Bob can postselect those cases in which P1 takes the long path and P2 takes the short path on the return trip. Thus the interfering amplitudes experience identical delays on their round trip, ensuring insensitivity to drift in Bob's interferometer.

The role of the phase modulators can be readily understood by examining the space-time diagram of this protocol [see Figure 10.4(B)]. The eight boxes (A1–A4, B1–B4) refer to the phase settings on the two modulators as the two amplitudes pass through each of them twice. For example, B2 refers to the phase acquired by the delayed amplitude of the pulse that Bob sends to Alice, while B4 refers to the phase acquired by the same amplitude as it travels back from Alice to Bob. It should be understood that B1–B4 refer to settings of the same physical phase shifter at different times (and similarly for A1–A4). The probability of a detection at Bob's detector is given by

$$P_d \propto 1 + \cos[(B2 - B1) + (A2 - A1) + (A4 - A3) + (B4 - B3)]. \quad (10.4)$$

From this expression we see that only the relative phase between the phase modulator settings affects the probability of detection. Thus, by setting $B1 = B2$ and $A1 = A2$, Alice and Bob can implement the interferometric version of BB84 [14] by encoding their cryptographic key in the difference settings $\Delta\phi_A \equiv A4 - A3$ and $\Delta\phi_B \equiv B4 - B3$. Since the resulting expression

$$P_d \propto 1 + \cos(\Delta\phi_A + \Delta\phi_B) \quad (10.5)$$

is independent of the time delay in Bob's interferometer and the absolute phase settings in either modulator, Alice and Bob are able to achieve high-visibility interference without initial calibration or active compensation of drift.

10.3.2   One-Way Noise-Immune Time-Bin-Coded QKD

In this section, we describe a one-way noise-immune time-bin-coded QKD scheme. The scheme also allows for Bob's apparatus to be passive. Before presenting the full scheme, we review a non-noise-immune QKD scheme that motivates the technique used to combine noise-immunity and passive detection.

The two-photon quantum key distribution scheme described in Reference [8] has the remarkable property that both Alice and Bob use passive detection (i.e., they are not required to switch between conjugate measurement bases). In Reference [2], Gisin et al. suggest applying the AWI to generate an associated one-photon scheme. We present a specific implementation of this one-photon scheme here to show that it achieves passive detection by enlarging the Hilbert space (see Figure 10.5). Let the advanced and delayed

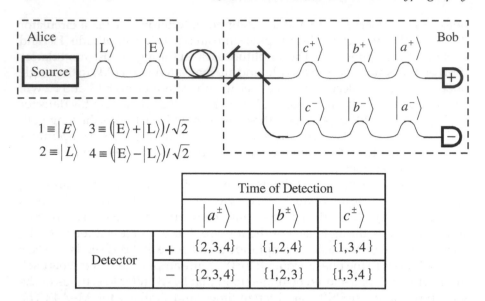

$1 \equiv |E\rangle$ $3 \equiv (|E\rangle + |L\rangle)/\sqrt{2}$

$2 \equiv |L\rangle$ $4 \equiv (|E\rangle - |L\rangle)/\sqrt{2}$

		Time of Detection		
		$|a^{\pm}\rangle$	$|b^{\pm}\rangle$	$|c^{\pm}\rangle$
Detector	+	{2,3,4}	{1,2,4}	{1,3,4}
	−	{2,3,4}	{1,2,3}	{1,3,4}

Figure 10.5   A single-photon implementation of BB84 suggested in Reference [2]. The kets $|E\rangle$ and $|L\rangle$ correspond respectively to an advanced (early) and a delayed (late) single-photon wavepacket. Alice sends one of the four states listed below the diagram of the apparatus. The chart indicates which of Alice's states are consistent with a given measurement event at Bob's side. As described in the text, Bob's apparatus does not require active change of measurement basis.

single-photon wavepackets be associated with the poles of the Poincaré sphere. The four states required for BB84 are typically taken from the equator, since a single Mach–Zehnder interferometer can be used to generate any of the equatorial states. Instead, we imagine using two antipodal points on the equator and the poles themselves. Bob analyzes the signal from Alice with a Mach–Zehnder interferometer, recording which detector fired (one of two possibilities) at which time (one of three possibilities). When Bob's detection is in the first or third time positions, he can reliably distinguish between the pole states based on the time of detection. When his detection is in the second time position, he can reliably distinguish between the equatorial states based on which detector fired. Thus Bob is no longer obliged to make an active change to his apparatus to effect the requisite change of basis [23].

To see how this passive detection is derived from enlargement of the Hilbert space, consider the quantum state of Alice's signal after Bob's Mach–Zehnder interferometer. Alice's four states of one qubit are mapped onto four mutually nonorthogonal states of a six-state quantum system (see Figure 10.5). Thus by mapping a two-state quantum system into a six-state quantum system, Bob is able to perform his part of the BB84 protocol with a fixed-basis measurement in the six-state Hilbert space [24].

Next we present a scheme that combines passive detection with one-way noise-immunity (see Figure 10.6). This scheme follows from that presented

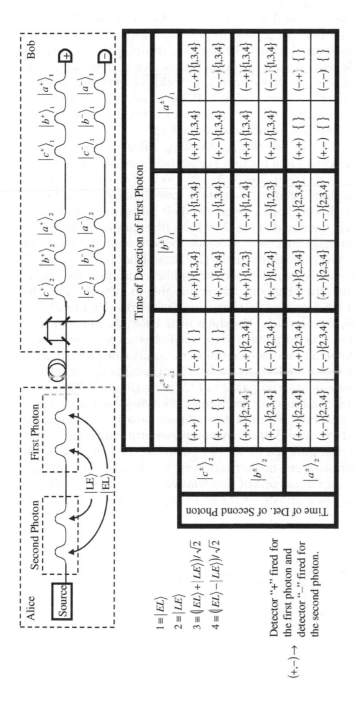

Figure 10.6 A schematic of one-way noise-immune time-bin-coded QKD [see Figure 10.1(E)]. Two time-bin qubits are sent from Alice to Bob in one of the four quantum states on the left of the figure. The chart on the right uses two levels of structure to describe the detection pattern at Bob's side. The coarse structure is defined by the bold lines. Each of the nine bold-lined rectangles corresponds to a specification of the joint time of detection of the two photons. The fine structure is defined by the thin lines. Each of the four thin-lined rectangles within a bold-lined rectangle corresponds to a specification of which detector fired for each of the two photons (this coding is illustrated by an example at the bottom left of the figure). The numbers in the curly brackets in each thin-lined rectangle indicate which (if any) of the four quantum states on the left are consistent with the corresponding detection pattern.

in Reference [6], just as the preceding single-photon scheme follows from the traditional phase-coding implementation. Let the states |1⟩ and |2⟩ in Figure 10.6 be associated with the poles of the Poincaré sphere. Instead of using equatorial states and forcing Bob to postselect those cases for which the advanced (delayed) amplitudes take the long (short) path, we use two equatorial points (|3⟩ and |4⟩) and the poles themselves to make up Alice's four signal states. Signal states that are consistent with a given joint detection are presented in the chart. As seen in Figure 10.5, each photon can lead to six different detection events. Thus, since the new protocol involves two photons, there are 36 possible detection events (see Figure 10.6).

The protocol operates as follows. As in BB84, Alice and Bob publicly agree on an association of each of the four signal states (see Figure 10.6) with logical values 0 or 1 (i.e., $1 \rightarrow 0$, $2 \rightarrow 1$, $3 \rightarrow 0$, $4 \rightarrow 1$). For each run of the experiment, Alice randomly chooses one of the four signal states and sends it to Bob. When Bob detects both photons in their respective middle time slots, he has effectively measured in the {3, 4} basis (the "phase" basis). When Bob detects both photons in their early time slots, or both photons in their late time slots, he has effectively measured in the {1, 2} basis (the "time" basis) [25]. After the quantum transmission, Alice and Bob publicly announce their bases. On the occasions when their bases match, Bob is able to infer the state that Alice sent, based on his detection pattern using the chart in Figure 10.6. As in single-qubit BB84, the occasions in which their bases do not match are discarded. The scheme achieves passive detection (Bob is not required to make any active changes to his apparatus) and noise-immunity (the phase delay in Bob's interferometer does not affect any measured probabilities). The intrinsic efficiency of the scheme is $1/4$, compared to $1/2$ for single-qubit BB84.

A proposed implementation for the source employed in Figure 10.6 is presented in Figure 10.7. First, a pair of noncollinear, polarization-entangled photons is produced via type-II spontaneous parametric down-conversion from a nonlinear crystal pumped by a brief pulse [26]. Second, the modulating element M performs one of four functions (filters one of the two polarization modes, or introduces one of two relative phases between the two polarization modes), based on Alice's choice of signal states. Third, the two

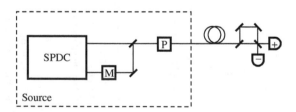

Figure 10.7 A proposed implementation for the source employed in Figure 10.6. SPDC is a nonlinear crystal pumped by a brief pulse to produce a noncollinear, polarization-entangled two-photon state via spontaneous parametric down-conversion. The action of elements M and P is described in the text.

beams are combined with a relative temporal delay that matches the temporal delay that Bob will subsequently introduce with his Mach–Zehnder interferometer. This stage converts the photon pair from a pair of spatially defined polarization-entangled qubits to a pair of polarization-defined time-bin-entangled qubits. Finally, the element labeled P (for polarization) delays and rotates one of the polarization modes by a duration much greater than the delay of the third step, so that the delayed portion of the state is in the same polarization mode as the nondelayed portion. Thus the two photons sent from Alice to Bob have the wavepacket structure illustrated at the top of Figure 10.6.

There are two noteworthy aspects of the configuration in Figure 10.7. First, the technique introduced in Reference [8] for creating time-bin-entangled photons pairs only leads to superpositions of the correlated possibilities (i.e., $|EE\rangle$ and $|LL\rangle$). The source presented in Figure 10.7 enables arbitrary superpositions of the anticorrelated possibilities (i.e., $|EL\rangle$ and $|LE\rangle$). Furthermore, the correlated states can easily be created from this source by rotating the polarization axes at element M in Figure 10.7. In this way, all four time-bin-entangled Bell states can be conveniently generated with this source. Second, the interference in Bob's interferometer results from the indistinguishability of photon amplitudes that were initially in the same polarization mode. This is in contrast to configurations in which photon amplitudes from different polarization modes are made indistinguishable by use of a polarization analyzer. Thus the reduction in visibility that has come to be associated with extremely brief pump pulses [15] will not be present in this scheme. Note that a symmetrization method has been developed to restore visibility for experiments using polarization-entangled photons created by such a short pulse pump [16,17].

10.3.3 Symmetric Noise-Immune Time-Bin-Coded QKD

In the symmetric time-bin scheme of Figure 10.1(F), the source produces a four-photon entangled state. As it is currently not practical to create such a state, we achieve the same result in Figure 10.8 by using two entangled pairs in the state

$$(|EE\rangle_{13} + |LL\rangle_{13})(|EE\rangle_{24} + |LL\rangle_{24}), \tag{10.6}$$

where E and L stand for early and late, respectively. The source apparatus consists of three switches, while Alice and Bob simply have Mach–Zehnder interferometers. The switches in the source behave as follows. The first switch (SW1) directs photon 1 along the lower path and photon 2 along the upper path. The action of the second switch (SW2) is indicated by the labels t and r, which stand for transmit and reflect, respectively. Thus for the early amplitude of photon 1 and the late amplitude of photon 2, SW2 reflects; otherwise it transmits. The third switch (SW3) directs the photons 5 and 6 onto the same output fiber. By postselecting only those occasions when one photon is found in the positions labeled 5 and 6, Alice effectively creates the

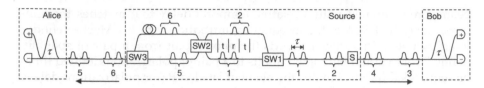

Figure 10.8 A schematic of symmetric noise-immune time-bin-coded QKD [see Figure 10.1(F)]. A central source (S) emits two separately entangled photon pairs [see Equation (10.6)]. One photon from each pair is sent to Bob. The other two photons are sent through a series of three switches. The first switch (SW1) directs photon 1 along the lower path and photon 2 along the upper path. The action of the second switch (SW2) is indicated by the labels t and r, which stand for transmit and reflect, respectively. The third switch (SW3) directs photons 5 and 6 onto the same output fiber. By postselecting the cases in which one photon is in position 5 and one photon is in position 6, Alice effectively creates the four-photon entangled state in Equation (10.7). This state is then analyzed by Alice and Bob with their Mach–Zehnder interferometers, each of which has a delay equal to τ. The protocol used to establish a shared key is described in the text.

four-photon entangled state

$$|ELLE\rangle_{5634} + |LEEL\rangle_{5634}. \tag{10.7}$$

When all the amplitudes follow the pattern ($E \rightarrow$ long path, $L \rightarrow$ short path) in Alice's and Bob's Mach–Zehnder interferometers, Alice and Bob announce that they have measured in the phase basis, and they use the chart in Figure 10.9 to infer the bit value. When one photon on each side does not

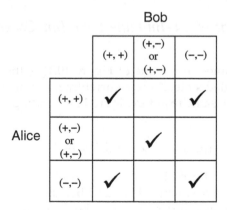

Figure 10.9 Possible joint detection patterns for the scheme of Figure 10.8. The expression $(+, -)$ indicates that the $+$ detector fired for the first photon and the $-$ detector for the second. Given that the source produces the state in Equation (10.6), when all the amplitudes follow the pattern ($E \rightarrow$ long, $L \rightarrow$ short) in Alice's and Bob's Mach–Zehnder interferometers, the unchecked joint detection patterns do not occur because of destructive interference. Thus Alice and Bob may use a publicly known encoding (e.g., $\{(+, +), (-, -)\} \rightarrow 0$, $\{(+, -), (-, +)\} \rightarrow 1$) to agree on a secret key bit.

follow the pattern ($E \to$ long, $L \to$ short), Alice and Bob announce that they have measured in the time basis. On these occasions, they each know which of the superposed terms in Equation (10.7) was realized, and they use this knowledge to establish a shared bit. The scheme is noise-immune because on the phase-basis occasions, each leg of the two Mach–Zehnder interferometers is traversed by one of the four photons. Thus the relative phase along the two paths of each interferometer factors out and does not affect the measured results. The scheme is passive because neither Alice nor Bob is required to make active changes to their apparatus.

The security of the scheme derives from the fact that only the state in Equation (10.6) will produce the correlations that Alice and Bob measure. Therefore the source can be controlled by the adversary without compromising security. This technique can be viewed as the time-bin analog of the polarization based entanglement distillation experiment described in Reference [18].

10.4   Discussion

We have presented round-trip, one-way, and symmetric noise-immune QKD schemes that can be implemented with existing technology for both polarization and time-bin qubits. The noise-immunity of the schemes makes active compensation of interferometric drift and channel birefringence unnecessary. The round-trip methods are the simplest, since they do not involve entanglement. However, the bidirectional flow of signals leaves an opportunity for an eavesdropper to compromise the security of the link by sending signals into the apparatus of Alice and/or Bob and measuring the state of the reflected signal. The one-way schemes remove this security concern at the cost of requiring a multi-photon entangled state. A further advantage of the one-way schemes presented here is that they do not require Bob to make active changes to his apparatus. Finally, the symmetric schemes presented here achieve noise-immunity while requiring neither Bob nor Alice to make active changes to his/her apparatus. The cost of this simplicity is a doubling of the number of photons involved in each run of the protocol.

It is interesting to observe that discoveries in the field of quantum information (entanglement swapping and entanglement distillation) can be naturally related to other areas of quantum information theory (quantum error correction and decoherence-free subpaces) via the AWI, as demonstrated in Figure 10.1. Since the central goal of quantum computation is a "folding in time" of a classical computation, the AWI may yield insight into the mechanisms behind the speed-up achieved by certain quantum computation algorithms.

References

1. M.A. Nielsen and I.L. Chuang, *Quantum Computing and Quantum Information*, Cambridge University Press, Cambridge, 2000.
2. N. Gisin, G. Ribordy, W. Tittel, and H. Zbinden, *Rev. Mod. Phys.*, 74, 145, 2002.

3. J.-C. Boileau, D. Gottesman, R. Laflamme, D. Poulin, and R. Spekkens, quant-ph/0306199, 2003.
4. A. Muller, T. Herzog, B. Huttner, W. Tittel, H. Zbinden, and N. Gisin, *Appl. Phys. Lett.*, 70, 793, 1997.
5. D.S. Bethune and W.P. Risk, *IQEC'98 Digest of Postdeadline Papers*, 12–2, 1998.
6. Z. Walton, A.F. Abouraddy, A.V. Sergienko, B.E.A. Saleh, and M.C. Teich, *Phys. Rev. A*, 67, 062309, 2003.
7. Z. Walton, A.F. Abouraddy, A.V. Sergienko, B.E.A. Saleh, and M.C. Teich, *Phys. Rev. Lett.*, 91, 087901, 2003.
8. J. Brendel, N. Gisin, W. Tittel, and H. Zbinden, *Phys. Rev. Lett.*, 82, 2594, 1999.
9. A.V. Belinsky and D.N. Klyshko, *Laser Phys.* (Moscow), 2, 112, 1992.
10. C.H. Bennett, *Phys. Rev. Lett.*, 68, 3121, 1992.
11. M. Martinelli, *Opt. Comm.*, 72, 341, 1989.
12. S.L. Braunstein and A. Mann, *Phys. Rev. A*, 51, R1727, 1995.
13. M. Zukowski, A. Zeilinger, M.A. Horne, and A.K. Ekert, *Phys. Rev. Lett.*, 71, 4287, 1993.
14. C.H. Bennett and G. Brassard, *Proceedings of the IEEE International Conference on Computers, Systems and Signal Processing*, 175–179, 1984.
15. T.E. Keller and M.H. Rubin, *Phys. Rev. A*, 56, 1534, 1997.
16. F. De Martini, G. Di Giuseppe, and S. Pádua, *Phys. Rev. Lett.*, 87, 150401, 2001.
17. Y.-H. Kim and W.P. Grice, *J. Mod. Optics*, 49, 2309, 2002.
18. T. Yamamoto, M. Koashi, S.K. Ozdemir, and N. Imoto, *Nature*, 421, 343, 2003.
19. H. Bechmann-Pasquinucci and W. Tittel, *Phys. Rev. A*, 61, 62308, 2000.
20. K. Inoue, E. Waks, and Y. Yamamoto, *Phys. Rev. Lett.*, 89, 37902, 2002.
21. W. Tittel, J. Brendel, H. Zbinden, and N. Gisin, *Phys. Rev. Lett.*, 84, 4737, 2000.
22. One can always convert an active-detection scheme to a passive-detection scheme by using a beam splitter to probabilistically send the received photon(s) to one of some number of separate detection setups. A drawback of this approach is that the number of optical elements required is increased. The passive schemes described in this chapter, like that in Reference [8], are "intrinsically passive," in that they achieve passive operation without increasing the number of optical elements required.
23. The idea of using pole states is explored in Reference [19]; however, that paper does not mention the possibility of passive detection.
24. A similar idea is presented in Reference [20]. In that paper, Alice uses four states of a three-state quantum system, and Bob achieves passive detection by mapping Alice's three-state quantum system into an eight-state quantum system.
25. On the occasions when Bob's detection pattern is (early, middle), (middle, early), (middle, late), or (late, middle), he has also effectively measured in the time basis. However, to simplify the analysis by making the probability of successful bit-sharing independent of the basis in which Alice sent, we consider only the extreme cases (early, early) and (late, late) as valid time-basis detections.
26. A femtosecond pump pulse is typically desired for experiments involving the simultaneous creation of multiple down-converted photon pairs [16]. Our implementation does not require such a brief pump pulse and will work with a picosecond laser, such as that used in Reference [21].

Index

A

Advanced wave interpretation, 212–213, 223
Alberti, Leone Battista, 2, 4
Ali, 85–86
Alice, 85–86, 92, 108, 117–118, 147, 164
Al-Kindi, 3
Ancilla photons, 131, 141
ARTEMIS-SPOT4 satellite-to-satellite link, 189, 201
Atmospheric turbulence, 202
Avalanche photo detectors, 90, 94
Average privacy amplification bound, 156–157

B

B92 protocol, 104, 112–113
Babbage, Charles, 3
β-Barium borate crystal, 66
BB84
    description of, 12, 23, 64, 103–104, 161
    prepare and measure strategy, 114
    protocol, 113, 160–161
    secrecy of, 146
    unconditional security of, 104
BBN key relay protocols, 99–100
BBN Mark 2 weak coherent system, 90–96
BBN quantum key distribution protocols, 96–97
Beam broadening, 202
Bell inequalities
    Clauser–Horne–Shimony–Holt, 29, 71
    description of, 28–30
    experiments with, 68

Bell state analyzer, 52, 58
Bell state measurement, 52
Bennett, Charles H., 13
Binary entropy function, 148
Blakley's secret sharing protocol, 166
Bob
    description of, 84, 87, 108, 114, 117, 164
    measurement systems, 119–120
Bohm, David, 8
Boris, 84

C

Caesar, Julius, 3
Caesar ciphers, 3–4
Carter-Wegman hash functions, 157
Ciphers
    Caesar, 3–4
    unbreakable, 5
Ciphertext, 4
Classical cryptography, 1
Clauser–Horne–Shimony–Holt Bell inequality, 29, 71
Clipper chip, 167–168
Cocks, C., 7
Coherent state cryptography
    description of, 109
    polarization encoding in, 110
Communication
    quantum, See Quantum communications
    secret, 2
Conditional NOT gate, 54
Continuous variable quantum cryptography, 105

Continuous variable quantum state
  sharing, 183
Continuous variable quantum
  teleportation, 169
Continuous variable systems, 114
Controlled NOT gate
  description of, 47, 131–132
  nondestructive, 58
  photonic, 55–60
  polarization-encoded qubits used
    with, 133
  schematic diagram of, 133
  for single photons, 138
Controlled phase operations, 55
Cryptanalysis, 5
Cryptographic keys, 7
Cryptography
  classical, 1
  coherent state, 109
  definition of, 164
  history of, 2–3, 5–6, 164
  origins of, 2–3
  public-key, 7
  quantum, See Quantum cryptography
  Shannon's contributions to, 5
  single-photon, 105

D

Danube experiment, 69–70
DARPA quantum network
  BBN Mark 2 weak coherent system,
    90–96
  connectivity schematic of, 85
  description of, 84
  future plans for, 100
  metro-fiber portions of, 85
  motivation for, 86
  status of, 84–86
  summary of, 100–101
Data protection, 188
de Vigenre, Blaise, 4
Defense Advanced Research Projects
  Agency, See DARPA quantum
  network
Dense wavelength division
  multiplexing, 94
Diffie–Hellman key exchange protocol, 7
Discrete variable systems, 114

E

Eavesdropping
  analyzing of, 11–12
  definition of, 8
  description of, 1–2
Effective secrecy capacity, 147–154
Einstein, Albert, 8
Einstein–Podolsky–Rosen entanglement,
  169
Ekert protocol, 29, 64
Electro-optic feedforward, 174
Electro-optic modulator, 118–119
Ellis, James, 7
Encryption
  quantum cryptography for generating
    material for, 146
  RSA system, 7–8
  substitution, 3
  transposition, 3
Energy-time entanglement quantum key
  distribution, 33
Enigma machine, 5
Entangled state key exchange, 198–199
Entangled state quantum cryptography,
  206–207
Entangled-photon approach, 130
Entanglement, See also Quantum
  entanglement
  BB84 scheme, 12, 23, 64
  definition of, 189–190
  description of, 62–64
  Einstein–Podolsky–Rosen, 169
  higher dimensional, 60–62
  protocols based on, 9–10, 105
  prototype system for, 64–67
  quantum key distribution, 32
  realizations, 27–34
  transmission of, over faulty channels,
    183–184
Entanglement purification
  description of, 46–47
  principles of, 54–55
  scheme for, 53
Entanglement swapping, 38, 46, 48–49
EPR program, 8
Error rates
  postselection effects on, 122
  quantum bit, 26, 85
Eve, 108, 113

Experimental cryptography
  polarization encoding, 106, 109–112
  postselection, 107–109
  protocol, 112–115

F

Faint coherent pulses
  description of, 104
  polarized, 192
Faint laser pulses, 24
Faint laser quantum key distribution,
  23–24
Faint pulse quantum cryptography
  eavesdropping susceptibility of, 206
  entangled state key exchange, 198–199
  key distribution systems, 204–206
  long-range trial of, 195–198
  method of, 190–191
  receiver for, 193–194
  satellite-to-ground, 207
  timing and synchronization for,
    194–195
  tools used in, 191–195
  transmitter for, 191–193
Faraday mirrors, 18, 130
Feedback loop, 128
Feedforward loop, reconstruction
    protocol using, 174–175
Fock basis, 107
4 + 2 state protocol, 104, 113–114
Four-photon quantum communication,
    36–39
Four-qubit encoding, 128, 141
Free-space distribution of quantum
    entanglement, 68–73
Free-space entangled-state distribution,
    198
Free-space links, 100, 188
Free-space optical links, 68–73
Free-space quantum cryptography
  description of, 188–189
  space applications of, 199–201, 207

G

Gaussian beam, 201–202
Gaussian modulated coherent states, 106
Gaussian white noise, 171
GHZ states, 168

H

Hash functions
  Carter-Wegman, 157
  description of, 155
Heisenberg uncertainty principle, 10,
    107
Hidden variable theory, 22
Hilbert space, 61, 170, 218
Homodyne detection, 105–106, 115,
    123

I

Interferometers
  advantages of, 129–130
  description of, 20, 23, 26, 33
  disadvantages of, 130
  Mach–Zehnder, 86–87, 92, 172,
    214–215, 217–218
  polarizing Sagnac, 139
  quantum key distribution based on,
    129–130
  two-photon, 129
Internet key exchange, 97

K

Kasiski, Friedrich Wilhelm, 3
Key
  definition of, 188
  secure distribution of, 188
Key distribution
  faint laser quantum, 23–24
  faint pulse quantum cryptography,
    204–206
  privacy amplification after, 154
  problems associated with, 6–8
  quantum, See Quantum key
    distribution
  security of, 12
Key exchange
  entangled state, 198–199
  quantum, 200–201
  to space, 201–207
Key relay
  benefits of, 89
  device for, 88
  for trusted networks, 89, 99–100

L

Le chiffre indéchiffrable, 3–4
LEO satellite, 200
Linear optics quantum logic gates
    description of, 131–135
    single-photon source and memory,
        135–139
Liu's problem, 165–166
Local realism, 8
Long-distance quantum correlation,
    29–32
Long-distance quantum teleportation,
    51–52
Lütkenhaus analysis, 153

M

Mach–Zehnder interferometer, 86–87,
    92, 172, 214–215, 217–218
Magneto-optic modulator, 118
"Man-in-the-middle" spoofing, 152
Mark 2 weak coherent system, 90–96
Mauborgne, Joseph, 5
Mode-locked setup, 173
Monoalphabetic ciphers, 3
Multipartite quantum cryptography, 184
Multiphoton pulses, 151

N

National Institute of Standards and
    Technology, 84
No-cloning theorem, 36
Noise-immune polarization-coded
    schemes
    one-way quantum key distribution,
        214–215
    round-trip quantum key distribution,
        212–214
    symmetric quantum key distribution,
        215–216
Noise-immune time-bin-coded schemes
    one-way quantum key distribution,
        217–221
    round-trip quantum key distribution,
        216–217
    symmetric quantum key distribution,
        221–223

Noisy quantum channel, 54
Nonconditional teleportation, 50
Nonorthogonality, 112
Nonunity photodiode efficiency, 121

O

One-time pads, 6
One-way noise-immune
    polarization-coded quantum key
    distribution, 214–215
One-way noise-immune time-bin-coded
    quantum key distribution,
    217–221
Optical parameter amplifiers, 172, 177
Optical photons, 68
Orbital angular momentum, 61–62

P

Parametric down-conversion, 132, 135,
    138
Peltier thermoelectric coolers, 95–96
Perfect secrecy, 155
Phase modulators, 217
Phase sensitive amplifiers, 174
Photon counting, 104, 123
Photon number splitting, 36, 92, 150
Photon pairs, 27, 46
Photonic controlled NOT gate, 55–60
Photonic switching, 89, 97–98
Plaintext, 4
Planck's constant, 110
Plug & play, 24–25
Plug-and-play quantum cryptography,
    216
Podolsky, Boris, 8
Poincaré sphere, 20, 23, 111, 113, 220
Pointwise privacy amplification bound,
    156, 158–160
Pointwise probability parameter, 157
Polarization correlation, 70–71
Polarization encoding, 106, 109–112
Polarization measurement systems,
    119–120
Polarization-coded schemes,
    noise-immune
    one-way quantum key distribution,
        214–215

round-trip quantum key distribution, 212–214
   symmetric quantum key distribution, 215–216
Polarization-entangled qubits, 53
Polarized coded faint pulses, 192
Polarized photons, 24–25
Polarizing beam splitter, 56–58, 132–134, 190
Polarizing Sagnac interferometer, 139
Polyalphabetic ciphers
   Alberti's discovery of, 4
   breaking of, 4–5
Polynomial interpolation, 166
Porta, Giovanni Battista Della, 4
Positive operator valued measurement, 113
Postselection
   description of, 109, 114
   error rates reduced by, 122
   experimental, 121–123
Prepare and measure protocols, 10–11
Privacy amplification
   after key distribution protocol, 154
   average bound, 156–157
   description of, 146, 155–156
   pointwise bound, 156, 158–160
   result of, 146
   summary of, 161
Privacy amplification subtraction parameter, 157
Proactive secret sharing, 167
Public-key systems, 7
Pulsed parametric down-conversion, 138
Pump photon, 21

Q

Q-switched setup, 173
Quadrature entangled, 171
Quantum bit error rate, 26, 85
Quantum coding, 189–190
Quantum communications
   advances in, 46
   challenges in, 128–131
   description of, 18, 23, 46, 127–128
   four-photon, 36–39
   in space, 73–75
   summary of, 75

   three-photon, 36–39
   widespread use of, 131
Quantum conjugate coding, 13
Quantum correlation, long-distance, 29–32
Quantum cryptography
   classic, 87
   classical cryptography vs., 1
   continuous variable, 105
   description of, 1–2, 23, 145, 188–189
   encryption material generated by, 146
   entangled state, 206–207
   entanglement-based, See Entanglement
   future of, 34–39
   industrial application of, 62
   multipartite, 184
   plug-and-play, 216
   satellite-based, 207–208
   secret communications obtained using, 154
   single-photon, 63
   summary of, 13
   two-photon, 27–34
Quantum dots, 20
Quantum efficiency, 95
Quantum encoder
   description of, 133
   schematic diagram of, 136
Quantum entanglement, See also Entanglement
   description of, 35–36, 46
   free-space distribution of, 68–73
   purifying of, 54–55
Quantum error correction, 55, 182–183
Quantum information carrier, 60
Quantum information networks, 181–182
Quantum information theory, 223
Quantum key distribution
   BBN protocols, 96–97
   classic, 86
   coherent polarization states for, 106
   commercialization of, 39
   description of, 83–84, 212
   distance-limited, 83
   endpoints, 88
   energy-time entanglement, 33

entanglement-based protocols, 9–10, 29, 32
error sources in, 139
evolution of, 128
experimental demonstration of, 128
faint laser, 23–24, 27
Gaussian modulated coherent states, 106
history of, 103–104
homodyne detection, 105–106
interferometric approach for, 129–130
links, 88
one-way noise-immune polarization-coded, 214–215
one-way noise-immune time-bin-coded, 217–221
plug & play setup, 23–26
postselection analysis applied to, 109
prepare and measure protocols, 10–11
prototype, 67
round-trip noise-immune polarization-coded, 212–214
round-trip noise-immune time-bin-coded schemes, 216–217
satellite-to-ground, 200
symmetric noise-immune polarization-coded, 212–214
symmetric noise-immune time-bin-coded, 217
two-photon, 217
Quantum key distribution network
benefits of, 89–90
cost savings of, 90
definition of, 86–88
description of, 84
photonic switching for "untrusted networks," 89, 97–98
robustness of, 90
trusted, 89, 99–100
untrusted, 89, 97–98
Quantum logic gates
description of, 127
linear optics, 131–135
Quantum money, 169
Quantum noise, 11
Quantum parity check, 132, 134, 136
Quantum physics, 8–9
Quantum privacy amplification, 12
Quantum relay, 37

Quantum repeaters
description of, 47, 51, 139
linear optical techniques and, 128
logic gates in, 55
Quantum secret sharing
description of, 33, 168
threshold, 169
Quantum state sharing
applications of, 181–184
characterization of, 176–177
continuous variable, 183
dealer protocol, 170–171
definition of, 164
description of, 168–170
efficacy of, 176
entanglement transmission over faulty channels enabled using, 183–184
experimental realization of, 177–181
implementation of, 170–177
reconstruction protocols, 172–177
schematic diagram of, 164
signal-to-noise transfer, 176
threshold, 179
Quantum teleportation
description of, 36–37, 47–48
Innsbruck experiment of, 50
long-distance, 51–52
nonconditional, 50
scalable, 48–51
Quantum theory, 7
Qubit
definition of, 19, 60
polarization-entangled, 53
teleported, 48, 50
time-bin, 19–23
Qunits, 60
Qutrits, 62

R

Rarity, John, 130
Receiver, for faint pulse quantum cryptography, 193–194
Rejewski, Marian, 5
Reverse reconciliation, 106
Rosen, Nathan, 8
Round-trip noise-immune polarization-coded quantum key distribution, 212–214

Round-trip noise-immune
time-bin-coded schemes quantum
key distribution, 216–217
RSA encryption system, 7–8

S

S_3 modulation, 118
Satellites, 199–201
Satellite-to-ground faint pulse quantum
cryptography, 207
Scalable teleportation, 48–51
Scherbius, Arthur, 5
Scytale, 2–3
Secrecy capacity
effective, 147–154
for keys of finite length, 153–154
Secret sharing
applications of, 167–168
classical, 165–168
electronic voting applications of,
168
proactive, 167
quantum domain translation of,
168–170
threshold, 164, 166–167
verifiable, 167
Secret sharing protocols
Blakley's, 166
description of, 164
Shamir's, 166–167
Security, 11–13
Shamir's secret sharing protocol,
166–167
Shannon, Claude, 5
Shannon entropy, 149, 155
Shor's algorithm, 182
Signal state, 118–119
Signal-to-noise ratio, 60
Signal-to-noise transfer, 176
Single-photon based realizations,
27, 32
Single-photon cryptography, 105
Single-photon pulses, 149
Single-photon quantum cryptography,
63
Space
free-space quantum cryptography
applications, 199–201, 207
key exchange to, 201–207

Spartans, 2–3
SPDC, See Spontaneous parametric
down conversion
Spontaneous parametric down
conversion, 21
Spontaneous parametric
down-conversion, 48–49
Stokes operators, 109–110, 113
Sub shot noise polarization, 115–117
Substitution ciphers, 3
Symmetric noise-immune
polarization-coded quantum key
distribution, 215–216
Symmetric noise-immune
time-bin-coded quantum key
distribution, 221–223

T

Telecom wavelengths, 18
Teleportation, quantum
description of, 36–37, 47–48
Innsbruck experiment of, 50
long-distance, 51–52
nonconditional, 50
quantum repeater and, 51
scalable, 48–51
Teleported qubits, 48, 50
Thermoelectric coolers, 95–96
Three-photon quantum communication,
36–39
Threshold quantum secret sharing,
169
Threshold quantum state sharing, 179
Threshold secret sharing, 164,
166–167
Time-bin qubits, 19–23, 27
Time-bin-coded schemes, noise-immune
one-way quantum key distribution,
217–221
round-trip quantum key distribution,
216–217
symmetric quantum key distribution,
221–223
Transmitter, for faint pulse quantum
cryptography, 191–193
Transposition, 3
Trithemius, Johannes, 4
Trusted networks, key relay protocols
for, 89, 99–100

Two-photon interferometers, 129
Two-photon quantum cryptography,
    27–34
Two-photon quantum key distribution,
    217

U

Unambiguous state discrimination
    attack, 149, 151
Unitary gain point, 175
Universal local oscillator, 177
Untrusted quantum key distribution
    network, 89

V

Verifiable secret sharing, 167
Vernam, Gilbert, 5
Vernam cipher material, 159

W

Wiesner, Stephen, 13
Wigner function, 177

Z

Zeno gates, 142

Milton Keynes UK
Ingram Content Group UK Ltd.
UKHW040104071024
449327UK00019B/815